PRAISE FOR *TERRA VIVA*

'Vandana Shiva has led an extraordinary life as a scientist/activist and leader of a global movement for food sovereignty, and she tells her remarkable story in this powerful new memoir. Filled with important information and history on the corporate theft of biodiversity, *Terra Viva* also tells the rich stories of the grassroots fight to take back sacred community knowledge and rights, in India and around the world. Just as our world would be a lesser place without Vandana Shiva, our literary heritage would also be diminished without this crucial book.'

– Maude Barlow, author of *Blue Covenant*,
Right Livelihood Award laureate

'Like her previous celebrated books, Vandana Shiva's *Terra Viva* shines with ecofeminist wisdom, passionate activism, and impeccable systemic thinking. Against the background of her devastating critique of reductionist science and predatory capitalism, she gives a thrilling account of the numerous grassroots movements she founded, cofounded, and supported; movements to protect trees, seeds, water, soil, and community – essential strands in the web of life. I warmly recommend this brilliant and deeply moving book.'

– Fritjof Capra, author of *The Tao of Physics*
and *Patterns of Connection*

'This book is a testament to the great work that Vandana has done for the food and ecological movements. It is a testament to how she encouraged the safeguarding of biodiversity and fought for the rights of women and farmers' communities.'

– Carlo Petrini, founder, Slow Food

'Vandana Shiva is the most brilliant woman I know; she is impressive, with a global knowledge of the environmental problems created, above all, by social disorders and the selfishness of short-term power. The awareness that her work and her actions provoke is sublime. This book is brilliant. It will bring you light. And answers.'

– Gilles-Éric Séralini, molecular biologist,
co-author of *The Monsanto Papers*

'Vandana Shiva's brilliant, discerning mind, granite character, poetic soul, and her galvanising skills as a grassroots organiser have won her place in the pantheon of the most effective and beloved leaders of the global environmental movement. Among her extraordinary gifts are her grasp of the enmeshment between human and natural history and her deep understanding of agricultural and forest ecology. Shiva is the world's most eloquent and persuasive champion of family farms, traditional agricultural practices, and against the evils of chemical and industrial farming and GMO crops. She documents, in this book, the insidious role chemical agriculture plays in diminishing food quality, soils, microbiomes, species diversity, human health, personal freedom, and in the corrosive impoverishment of the human soul. She shows how chemical warfare on agricultural pests gives rise to totalitarian political systems and the commodification of both humanity and creation.'

– Robert F. Kennedy Jr.

'Vandana Shiva's life story is the living story of the struggle to save soil, seeds, and human sanity. This beautifully written memoir is a compelling narrative which should be read by everyone who loves land, life, and the integrity of our precious Planet Earth. Every page of this book is filled with passion and compassion. This is one of the most informative and inspiring books for all activists, present and future, who wish to see ecological, social, and spiritual transformation emerging out of our current chaotic predicament.'

– Satish Kumar, founder, Schumacher College

'In *Terra Viva*, Vandana Shiva retraces her life as an accomplished academic and savvy activist, with a deep understanding of the planet's ecosystem functions and the destructive impact of the Poison Cartel and its supporters. The optimist Dr Shiva lives by example. Reading this book will open your mind to the causes of the deep problems we have been pushed into by the actions of a few, and give you hope and courage to reverse the downward spiral.'

– Hans R. Herren, founder and president,
Biovision Foundation for Ecological Development

'A journey through Vandana's life and work is most inspiring. She is a phenomenal woman – her work and conviction on the issues she has confronted with such determination and courage are not only admirable but breathtakingly amazing. From tackling the water barons and the manufacturers of genetically modified seeds, to the pharmaceutical companies and the food markets of the world, she has shown such superior insight and knowledge that she has been able to floor these giants. In South Africa she has many followers, admirers, and devotees among whom I, too, am present. This book is a must for all of us who are interested in an alternative world order, where issues of environmental protection, conservation, food sovereignty, peace, and nonviolence are uppermost.'

– Ela Gandhi,
former member of parliament, South Africa

'*Terra Viva* is one of the most remarkable books I have ever read. It is an awe-inspiring tale of multiple struggles against multiple injustices against people and our environment; an engaging life story of how one perceptive individual can empower various groups of people to take on governments and corrupt corporations and win justice for them – a classic David versus Goliath. This book is a must because Dr Shiva's multiple successes give us hope that we can regenerate our planet, empower our people, and build fair societies that enable well-being for all of us. Her life story shows that she is one of our most extraordinary leaders like Mahatma Gandhi and Dr Martin Luther King.'

– André Leu,
international director, Regeneration International

'*Terra Viva* captures the dynamic vibrance of Mother Earth and the necessary practice of Right Livelihood. Vandana Shiva shares them through brilliant research, writing, speaking skills and also, importantly, getting real things done on the ground. She shows courage, competence and conscience and always speaks truth to power, assertively and consistently, without fear or favour. Her wisdom flows through this book, it shines through like multiple beacons of a lighthouse that will inspire more of us "glocally" to take up the many challenges of these dismal times.'

– Anwar Fazal, Multiversity International

'In *Terra Viva*, Vandana Shiva offers us rare insights into who she is and what has shaped her outlook and focus in life. The book chronicles her thoughts, knowledge, convictions, and a sustained commitment to movement-building and maintenance of life. Diversity, symbiosis, mutuality, and reciprocity are the pillars on which her work has been built. Thank you, Vandana, for the generous offering of this testament, this living book, this sharing of your life.'

– Nnimmo Bassey, author of *To Cook a Continent*

ALSO BY VANDANA SHIVA

Terra Viva
My Life in a Biodiversity of Movements

VANDANA SHIVA

Chelsea Green Publishing
White River Junction, Vermont
London, UK

Terra Viva was originally published in India in 2022 by Women Unlimited (an
associate of Kali for Women), 7/10, First Floor, Sarvapriya, Vihar, New Delhi, 110 016.

This edition published by Chelsea Green Publishing, 2022.

Printed in the United States of America.
First printing September 2022.
10 9 8 7 6 5 4 3 2 1   22 23 24 25 26

ISBN 978-1-64502-188-9 (paperback) | ISBN 978-1-64502-189-6 (ebook)
| ISBN 978-1-64502-190-2 (audio book)

Library of Congress Cataloging-in-Publication Data is available upon request.

Chelsea Green Publishing
85 North Main Street, Suite 120
White River Junction, Vermont USA

Somerset House
London, UK

www.chelseagreen.com

Contents

Looking
Forward,
Looking
Back

I WAS BORN IN THE DOON VALLEY IN 1952, TO A FATHER WHO had become a forest conservator after leaving the army and a mother who had become a farmer after leaving a senior government job in education. My parents had met during the war, and when my father proposed to my mother she agreed to marry him if he eventually left the army and if she could continue to work. They also decided to give up their caste name as part of the anti-caste movement, which was a very significant part of our independence struggle, and adopted the caste-neutral Shiva. Mother was posted in what became Pakistan after the tragic partition of India in 1947; she survived miraculously, but she had become a refugee. Refugees of Partition were rehabilitated – shopkeepers got shops, employees got jobs, farmers got land. Instead of taking up a government job equivalent to what she had lost, my mother decided to get rehabilitated as a farmer.

I was born five years after Partition, and my childhood was shaped by the forests in the Himalaya where my father was posted and by my mother's farm in the foothills. Nature was my first inspiration – and the study of nature my first passion – which is how I ended up becoming a physicist.

My most intimate memories of childhood are the sights and sounds, tastes and smells of the Himalayan forests where I grew up; they became my physical and intellectual cradle. I feel a deep umbilical connection to forests of rhododendron, oak and deodar, and to mountain streams. We lived in Chakrata when I was born and later moved to Nainital, Pithoragarh,

Tehri, Uttarkashi and Dehradun, which my parents decided to make their home. Today, these Himalayan regions are an independent state called Uttarakhand (the mountain state).

The British had annexed several Himalayan districts in 1815, mainly to exploit their forest wealth. Pine (called *chir*, locally) was felled on a large scale to make 'sleepers' for railway lines. The entire catchment of the river Ganges falls in my region; in the Garhwal Himalaya, an Englishman, Mr Wilson, obtained a lease in 1850 to exploit all the forests of the Bhagirathi river valley for a low annual rental of Rs 400. Under his axe, several valuable deodar and *chir* forests were clear-felled and completely destroyed. In 1864, inspired by Mr Wilson's flourishing timber business, British rulers of the north-western provinces obtained a lease for twenty years and engaged him to exploit these forests as well for them. European settlements, such as Mussoorie, created new pressures for the cultivation of food crops, leading to a large-scale felling of oak forests. Inspired by the economic success of Mr Wilson and the government, the Tehri state took over the management of forests in 1895. Between 1897 and 1899, forest areas were reserved and restrictions were imposed on village use; these restrictions were resented and completely disregarded by the villagers, which led to incidents of organised resistance against the authorities. On March 31, 1905, in response to the resistance, a Durbar Circular (No. 11) from the Tehri king announced modifications to these restrictions.

The contradictions between people's basic needs and the state's revenue requirements, however, remained unresolved, and in due course these contradictions intensified. In 1930, the people of Garhwal launched the non-cooperation movement to draw attention to the issue of forest resources; forest satyagrahas to resist oppressive forest laws were most intense in the Rawain region. The king of Tehri was in Europe at that time; in his absence, Dewan Chakradhar Jayal resorted to armed intervention to crush a peaceful satyagraha at Tilari. A large number of unarmed satyagrahis were killed and wounded, while others lost their lives in a desperate attempt to cross the rapids of the Yamuna river. Years later, the martyrs of the Tilari massacre provided inspiration to the Chipko movement, when people pledged themselves to protect their forests.

There were very few roads in the Himalaya when I was growing up, so most of our journeys took place on foot or on horseback. As a forest conservationist, my father's job was to inspect and manage forests and regenerate them. During all our vacations we joined him on his tours. Our 'rations' would travel in huge boxes laden on mules, and there would always be another box full of books. We lived like nomads, far away from cities, amidst the wealth of the forest. This experience has clearly influenced my thinking about wealth and poverty; for me, the forests of my childhood were the source of abundance and beauty, diversity and peace. With my sister, I would collect ferns to turn into works of art; wildflowers were our pearls and diamonds. That is why when the forests started to disappear, I joined the Chipko movement to protect them.

The Chipko movement is historically, philosophically and organisationally an extension of the traditional Gandhian satyagraha. Its special significance lies in the fact that it took

place in post–Independence India; the continuity between
the pre–Independence and post–Independence forms of this
satyagraha was provided by Gandhians, including Sri Dev
Suman,[1] Mira Behn[2] and Sarala Behn.[3] Equipped with the
Gandhian worldview of development based on justice and
ecological stability, they contributed silently to the growth of
woman power and ecological consciousness in the hill areas
of Uttar Pradesh. The influence of Mira Behn and Sarala
Behn, the two European disciples of Gandhi, on the struggle
for social justice and ecological stability in the hills of Uttar
Pradesh has been immense – they generated a new breed of
Gandhian activists who laid the foundation for the Chipko
movement. Sunderlal Bahuguna[4] is prominent among this
new generation, deeply inspired by Mira Behn and Sarala
Behn. In an article written in 1952, Mira Behn had stated
that there was 'Something Wrong in the Himalaya':

> Year after year the floods in the north of India seem to
> be getting worse, and this year they have been absolutely
> devastating. This means that there is something radically
> wrong in the Himalayas, and that 'something' is, without
> doubt, connected with the forests. It is not, I believe, just
> a matter of deforestation as some people think, but largely
> a matter of change of species. Living in the Himalayas,
> as I have been continuously now for several years, I have
> become painfully aware of a vital change in the species of
> trees which is creeping up and up the southern slopes –
> those very slopes which let down the flood waters onto
> the plains below. This deadly change-over is from *banj*
> (Himalayan oak) to *chir* pine. It is going on at an alarming
> speed, and because it is not a matter of deforestation but of
> change from one kind of forest to another, it is not taken

sufficiently seriously. In fact, the quasi-commercial forest department is inclined to shut its eyes to the phenomenon because the *banj* brings in no cash for the coffers, whereas the *chir* pine is very profitable, yielding as it does both timber and resin.

Mira Behn had identified not merely deforestation but change in species suitable to commercial forestry as the reason for ecological degradation in the Himalayas. She recognised that the leaf litter of oak forests was the primary mechanism for water conservation in mountain watersheds.

Mira Behn and Sarala Behn were frequent visitors to our home. Sunderlal Bahuguna and Bimla Bahuguna,[5] too, came to meet my parents, and Ghanshyam Raturi (Shailani), the legendary Chipko poet, would spend hours with my mother, reciting his new songs. Our home was open house for social activists, poets and intellectuals; its stimulating environment must have been part of the informal education that made ecological values and values of social and economic equality basic to my life and work.

In 1972, women in a high-altitude village, Reni, blocked logging operations by wrapping their arms around the trees, giving birth to the Chipko (literally, to cling) movement. The name was given to the movement by Raturi, who composed folk songs which were sung by every child, woman and man in Garhwal.

Nineteen seventy-two witnessed the most widespread organised protests against commercial exploitation of Himalayan forests by outside contractors – in Uttarkashi on December 12, and in Gopeshwar on December 15. It was during these two protest meetings that Raturi composed his famous poem describing the method of embracing the trees to save them from being felled:

> *Embrace the trees and*
> *Save them from being felled;*
> *The property of our hills,*
> *Save it from being looted.*

In 1973, the tempo of the movement in Uttarkashi and Gopeshwar reached new heights. Raturi and Chandi Prasad Bhatt[6] were its main organisers; while a meeting of the Sarvodaya Mandal was in progress in Gopeshwar in April 1973, the first popular action to chase away the contractors erupted spontaneously in the region when villagers demonstrated against the felling of ash trees in Mandal forest. Bahuguna immediately asked his colleagues to proceed on foot to Chamoli district, following the axe-men and encouraging people to oppose them wherever they went. Later, in December 1973, there was a militant non-violent demonstration in Uttarkashi in which thousands of people participated. In March 1974, twenty-seven women under the leadership of 50-year-old Goura Devi[7] saved a large number of trees from a contractor's axe in Reni, a village that lies on the road from Joshimath to Niti Ghati, following which the government was forced to abolish the private contract system of felling. This was the first major achievement of the movement and marked the end of one phase.

During the next five years, Chipko resistance spread to various parts of the Garhwal Himalaya. It is important to note that it was no longer the old demand for the supply of forest products to local small industries, but a new one for ecological control of forest resource extraction to ensure a supply of water and fodder that was being aired. Among the numerous examples of Chipko's successes throughout the Garhwal Himalaya in the later years are those in the Adwani, Amarsar and Badiyargarh forests. The Adwani forests were scheduled to be felled in the first week of December 1977. Large groups of women led by Bachhni Devi came forward to save the trees. (Interestingly, Bachhni Devi was the wife of the local village head, who was himself a contractor.) Chipko activist Dhoom Singh Negi[8] supported the women's struggle by undertaking a fast in the forest; women tied sacred threads to the trees, symbolising their vow of protection. Between December 13 and 20, a large number of women from fifteen villages guarded the forests, while discourses from ancient texts on the role of forests in Indian life continued non-stop. It was here in Adwani that the ecological slogan 'What do the forests bear? Soil, water and pure air' was born.

The axe-men withdrew, only to return on February 1, 1978, with two truckloads of armed police. The plan was to encircle the forests with the help of the police, in order to keep the people out during the felling operation. Even before the police reached the area, volunteers of the movement entered the forest and explained their case to the forest labourers who had been brought in from distant places. By the time the contractors arrived, each tree was being embraced by three volunteers. The police, seeing the level of awareness among the people, hastily withdrew before nightfall.

In March 1978, a new auction was planned in Narendranagar. A large popular demonstration was organised against it and the police arrested twenty-three Chipko volunteers, including women. In December 1978, a massive felling programme was planned by the public sector undertaking, the Uttar Pradesh Forest Corporation, in the Badiyargarh region. The local people immediately informed Bahuguna who began a fast unto death at the felling site in January 1979. On the eleventh day of his fast he was arrested in the middle of the night; this only served to strengthen the commitment of the people. Ghanshyam Raturi and a priest, Khima Shastri, led the movement as thousands of men and women from neighbouring villages joined them in the Badiyargarh forests. The people guarded the trees for eleven days, after which the contractors finally withdrew. Bahuguna was released from jail on January 31, 1979.

The cumulative impact of sustained grassroots struggles to protect forests resulted in a rethinking of forest management in the hill areas. The Chipko movement's demand for declaring the Himalayan forests as 'protection' forests instead of 'production' forests for commercial exploitation was recognised at the highest policy-making level. The then prime minister, Indira Gandhi, after a meeting with Bahuguna, recommended a fifteen-year ban on commercial green felling in the Himalayan forests of Uttar Pradesh.

The moratorium on green felling gave the Chipko movement breathing time to expand its base, and Bahuguna undertook a 4,780 km long, arduous march from Kashmir to Kohima in Nagaland, contacting villagers along the long Himalayan range and spreading the message of Chipko. At the same time, activists found it opportune to take the movement to other mountain regions in the country.

I decided in 1974 that, while pursuing my PhD in quantum theory, I would volunteer with Chipko every vacation. And that is what I did.

Chipko was clearly my university for ecology. While my parents provided the embedding in a forest culture and an appreciation of natural mixed forests, it was Chipko that made me realise, in intimate detail, how biodiversity is at the heart of sustainable economies and how nature provides for the basic needs of the large majority in the world. As I worked with peasant women, transferring fertility from the forest to the field, I learnt my first lessons in organic farming: sustainable societies rest on humus. In those early years, as I moved between quantum physics and protecting the Himalayan forests, I learnt to respect both the best of modern ecological science and the best of traditional knowledge. I developed a humility about my doctorate studies, recognising how much I did not know, and how much knowledge illiterate village women, with no formal education, had. That is why the term 'knowledge society' as a description of computer-based societies is so inaccurate and misleading to me, implying that non-industrialised, non-computerised societies are without knowledge. In the case of biodiversity, of forest species and plant species, this is clearly not true; women and indigenous communities, the excluded of the industrial world, are the real custodians of biodiversity-related knowledge.

Nuclear physics was my chosen speciality until I realised that the science had a dark side to it. I changed course to become a theoretical physicist and worked in foundations of quantum theory, fully expecting to stay on and become a professor, when I was confronted with the nagging thought that I wasn't informed enough about how society works. We in India have the third largest scientific community in the world. We are among the poorest of countries. Science and technology are supposed to boost growth, remove poverty. So, where is the gap? I wanted to answer this question for myself, so I took three years off to look at science policy issues, be a little more educated socially, and then go back to physics. I went to the Indian Institute of Science and the Indian Institute of Management (IIM) in Bangalore, where I studied interdisciplinary research in science, technology and environmental policy.

After three or four years, what started as a look at policy issues became the focus of my life. Speeding me along a path that would ultimately lead to grassroots activism was my growing reputation as an authority in the field of environmental impact. I found myself learning more and more about the threat posed by biotechnology to biodiversity. In 1981, the Ministry of Environment invited me to study the effects of mining in the Doon valley. As a result of my report, the Supreme Court banned mining there in 1983. This was the first time I had done something professionally about conservation – it was not just an academic engagement divorced from action or consequences; I found it so fulfilling, working with communities and making a difference in society.

Research by itself cannot save the environment. Empowered communities are where action takes place, so I set up the Research Foundation for Science, Technology and Ecology (RFSTE) in Dehradun in 1981–82 to connect to communities and treat them as experts. I also decided on adopting a holistic approach to research because I believed that, for instance, geology cannot tell us that we are destroying water resources, geohydrology does that. It was this aspect of my work that was recognised when I was given the Right Livelihood Award in 1993 for creating a new paradigm for research and for working in novel ways with communities.

I realised during the great drought of 1984 in Karnataka that the very way we do agriculture is flawed. That year also saw the peak of militancy in Punjab. I wrote about the violence of the Green Revolution that gave rise to a non-sustainable form of agriculture, which was pretending to create more food but was actually destroying nature, farmers' sense of self, and creating war within society. What was really a sustainability and democracy issue was politicised and communalised.

In 1987, during a meeting at the United Nations (UN), I thought, Mahatma Gandhi had used a charkha or spinning-wheel to spearhead satyagraha; I came up with the idea of the seed as an equivalent of the charkha for our modern satyagraha against the appropriation of agriculture by multinational corporations (MNCs). Navdanya was born in that moment of awareness, although it did not become a full-fledged organisation until 1991. The conservation farm was started about five years later with a view to inspire farmers to come and see two-hundred-and-fifty varieties of rice and eight-hundred species of plants growing in the same field.

When Satish Kumar, editor of *Resurgence*, asked me to set up something in India along the lines of Schumacher College, which he had co-founded in the UK, I hesitated because I preferred building movements to building buildings. But he convinced me that it was time for an institution like the Bija (seed) Vidyapeeth, and so we set up the Bija Vidyapeeth/Earth University in 2000 – mainly because it was going to be at the Navdanya farm, which was a seed bank, and also because a seed is an inspiration for renewal as well as an example of the small embodying the whole. The Bija Vidyapeeth really became a bija; instead of buildings, I could see the progression of a dialogue and mutual growth. And so we continued to hold courses with Schumacher College, and the best of people agreed to come and teach: the physicist Fritjof Capra; The Body Shop founder, Anita Roddick; and Satish himself.

The farm at Pirumadara

My mother's farm was at Pirumadara (which means the place where a great pir rests), very close to the Jim Corbett National Park, a famous tiger reserve in the foothills of the Nainital Tarai. My older brother and sister were born in Nainital, and until I was twelve, I studied in a convent there. In the early years, weekends often meant going down to the farm at Pirumadara. We would first take a bus to the plains, and then a train. A bullock-cart would come to receive us at the station to take us to the farm, which was five miles away. A giant banyan tree dropped its twenty aerial roots down to create a bigger shelter than our little cottage on the farm, which had first been a tent, then a straw hut, and finally a small brick house. Most of my time was spent in the shade of the banyan, either sitting or playing on the swing. Our parents had planted an orchard and a garden, and every fruit imaginable was available for plucking – litchee and mango, guava and lemon, pomegranate and custard apple.

Next to the house flowed an irrigation canal, called 'gul' in the local language, where we would jump in and swim whenever we wanted to, and at mealtimes we would go off to the fields to collect mustard leaves or chickpea or bathua for saag – the typical broth of greens in Indian cuisine, nutritionally very rich in iron, vitamin A and B complex.

Thus, even though my parents were educated and middle class and we were sent to the fanciest English schools, we were also educated in ways that rural children are – by the trees and the soil. This is probably why ecology for me is not just an intellectual response; it is, in a very physical sense, the feel of the earth, and I am grateful to my parents for giving us the opportunity to grow up in the two worlds that India encompasses – a rural majority and a westernised elite. Though having access to a privileged education, we did not make it the source of arrogance, insulation and exclusion. It is this growing insulation between a rural India of farmers and an urban westernised elite that is dividing the country so deeply today and threatening her very survival as an equitable and ecological civilisation.

*****

When I was about four, we had to rush off to a small village called Duhai, near Meerut, where my mother's father had built the first school for rural girls. He had wanted the school to become a college. As was typical in those days, anyone working for social change used fasting as a means to draw attention to their cause. My grandfather not only did not eat any food, he also did not drink any water. He died fasting for girls' education. The President of India arrived one day after his death to announce that the school would now be a college.

My mother was the first in her community to have become a graduate. She studied in Lahore at Lady Maclagan College, and her mentor was Sir Chhotu Ram, who played an extremely important role in protecting the rights of peasants. The laws on land alienation that he saw passed in 1934 and 1936 were critical to preventing the creation of landlessness in Punjab and Uttar Pradesh, of the kind found in Bengal, Orissa and Bihar, where the British had introduced the zamindari system. In north-west India, land stayed in the hands of owner-cultivators, largely due to Chhotu Ram's efforts.

My mother was also deeply influenced by Gandhi, and travelled alone from Lahore to Poona to meet him when he was in jail. Her book *Two Words with My Sisters and Gandhi's Teachings* is based on the special brand of Indian feminism which can be called Gandhian socialist feminism; it is not a liberation based on an individual as an atomised construct, but the broad and deep emancipation of all members of society from all oppressions – social, cultural and economic.

My parents were the ultimate feminists – my father would cook and sew our clothes, my mother attended public meetings and joined politics. The subtle education in our home was that there was no difference between being a man or woman, boy or girl. This education in equality ran so deep that my sister and I would use the masculine pronoun for ourselves because we were copying our brother, who was the oldest. We grew up happily unaware of all these differences, and our parents did not consider it important to teach us gender discrimination. We were even allowed to choose our own names. I settled for Vandana when I was eleven; Mira, my sister, took equally long to adopt a name to her liking, but Kuldip, our brother, stayed with the name given to him by our parents.

We grew up as totally free spirits – seeking our personhood, unencumbered by the burdens of caste, religion and gender identity, and our parents took every care to create the context for that freedom that I hold so dear and defend so ferociously in everything I do.

More than half a century has passed since my childhood. My parents are no more, but Mira, Kuldip and I still share one world and each of us, in our own way, continues to live the values of simplicity and sharing, compassion and caring, conservation and protection that we imbibed from our wonderful parents.

*****

Physics was my passion and my chosen profession. In school I received the science talent scholarship, which gave me the opportunity to train in India's leading scientific institutions. I trained as a nuclear physicist at the Bhabha Atomic Research Centre in Bombay, but moved to theoretical physics when my sister, Mira, a medical doctor, made me aware of nuclear hazards. I realised then that most science is partial; I wanted to practice a holistic science and was drawn to quantum theory for its non-reductionist, non-mechanist paradigm.

Before leaving for Canada to do my PhD I wanted to visit my favourite places in the Himalayas, but the forests and streams had disappeared in the insane rush to build dams and roads, to grow apples by cutting down rich oak forests which absorb the monsoon rains to release the water slowly as streams.

I returned to India after my PhD because I wanted both to give back to my society and to understand it better, and so I chose the difficult and challenging path of trying to combine scientific research with social and ecological responsibility.

It was becoming increasingly evident to me that scientific expertise worked more in the service of capital and the destruction of nature, whereas I wanted to work in the service of people and nature. In 1981, I left academia and started the RFSTE to support grassroots ecology movements.

In 1984, a number of tragic events took place in India. In June, the Golden Temple was attacked because it was harbouring terrorists; by November, Indira Gandhi had been assassinated; and in December, the worst industrial disaster took place in Bhopal, when Union Carbide's pesticide plant leaked a toxic gas into the environment. Thirty thousand people died at the height of 'terrorism' in Punjab, thirty thousand died in the 'industrial terrorism' of Bhopal. This is equal to twelve 9/11s. I was forced to sit up and ask why agriculture had become like war. Why did the Green Revolution, which received the Nobel Peace Prize, breed extremism and terrorism in Punjab? This questioning led to my books *The Violence of the Green Revolution* and *Monocultures of the Mind*. Blindness to diversity and self-organisation in nature and society was clearly a basic problem in the mechanistic, Cartesian, industrial paradigm, and this blindness led to false claims that industrial monocultures in forestry, farming, fisheries and animal husbandry produced more food and were necessary to remove global hunger and poverty. On the contrary, monocultures produce less and use more inputs, thus destroying the environment and impoverishing people.

In 1987, the Dag Hammarskjöld Foundation organised a meeting in Geneva on biotechnology called 'Laws of Life'. At the conference, the biotech industry laid out its plans – to patent life; to genetically engineer seeds, crops and life forms; and to get full freedom to trade through the General Agreements on Tariffs and Trade (GATT) negotiations, which culminated in the creation of the World Trade Organization (WTO). This led to my focus on intellectual property rights, free trade, globalisation – and to a life dedicated to saving seeds and promoting organic farming as an alternative to a world dictated and controlled by corporations.

Having dedicated my life to the defence of the intrinsic worth of all species, the idea of life forms, seeds and biodiversity being reduced to corporate inventions and, hence, corporate property, was abhorrent to me. Further, if seeds become 'intellectual property', saving and sharing them becomes an intellectual property theft! Our highest duty, to save seeds, becomes a criminal act. The legalising of the criminal act of owning and monopolising life through patents on seeds and plants was morally and ethically unacceptable to me. So I started Navdanya, which promotes biodiversity conservation, as well as seed-saving and seed-sharing among farmers. An earth-centred, women-centred movement,

Navdanya has created over a hundred community seed banks through which seeds are saved and freely exchanged among our 300,000 members. We have brought back forgotten foods like jhangora (barnyard millet), ragi (finger millet), marsha (amaranth), naurangi dal and gahat dal. Not only are these crops more nutritious than globally traded commodities, they are more resource-prudent, requiring only 200–300 mm of rain compared to 2,500 mm for chemical rice farming. Millets could increase food production four hundred-fold, using the same amount of limited water. These forgotten foods are foods of the future, and farmers' seeds are the seeds of the future. For the farmer, the seed is not merely the source of future plants and food; it is the storage place of culture, of history. Seed is the first link in the food chain; it is the ultimate symbol of food security.

The free exchange of seed among farmers has been the means for maintaining biodiversity as well as food security, and is based on cooperation and reciprocity. A farmer who wants to exchange seed generally gives an equal quantity of seed from his field in return for the seed he gets. But this exchange goes beyond seeds; it involves an exchange of ideas and knowledge, of culture and heritage. It is an accumulation of tradition, of knowledge, of how to work the seed. Farmers gather knowledge about the seeds they want to grow by watching them grow in other farmers' fields, by learning about drought and disease and pest resistance.

In saving seeds and biodiversity we are protecting cultural diversity. Navdanya means nine seeds; it also means 'new gift' in the face of the extinction of species and the extinction of small farmers. The nine seeds and their respective navgrahas (nine cosmic correlates) are: Yava (barley) represents the sun; Shamaka (little millet) represents the moon; Togari (pigeon-pea) represents Mars, which is responsible for controlling the nervous system; Madga (moong) represents Mercury and stimulates intelligence; Kadale (chickpea) represents Jupiter; Tandula (rice) represents Venus; Til (sesame) represents Saturn and is characterised by oil; Maasha (black gram) represents Rahu; Kulittha (horse gram) represents Ketu.

Biodiverse systems produce more food and higher incomes than industrial monocultures. Our baranaja (twelve seeds) system yields twice as much output and three times higher incomes than a monoculture of corn. The twelve crops are: Phapra (*Fagopyrum tataricum*); Mandua (*Eleusine coracana*); Marsha (*Amaranthus frumentaceous*); Bhat (*Glycine soja*); Lobia (*Vigna catjang*); Moong (*Phaseolus mungo*); Gahat (*Dolichos biflorus*); Rajma (*Phaseolus vulgaris*); Jakhia (*Cleome viscosa*); Naurangi (*Vigna umbellata*); Jowar (*Sorghum vulgare*); and Urad (*Vigna mungo*).

Our conservation of heritage rice varieties has led to the protection of the original, authentic basmati as part of a slow food presidium. We have saved more than three thousand rice varieties, including over thirty aromatic varieties. The saline resistant seeds we have saved helped farmers in Orissa recover from the super cyclone of 1999 which killed thirty thousand people. These seeds were also distributed by Navdanya in rehabilitation after the devastating 2005 tsunami that hit the south east coast of India. We are creating 'Seeds of Hope' seed banks to deal with climate chaos. Heritage seeds that can survive droughts, floods and cyclones are collected, saved, multiplied and distributed. Farmers' seed-breeding is far ahead of scientific breeding and genetic engineering in providing flood-resistant, drought-resistant, and saline-resistant varieties. In the context of farmers' heritage, genetic engineering is, in fact, a laggard technology.

Not only are corporate, industrial breeding strategies incapable of dealing with climate change, genetically engineered seeds are killing farmers. In India, lakhs of farmers have committed suicide because of debt caused by high costs and unreliable seeds sold by corporations. Suicides are concentrated in areas which have become dependent on commercial seeds, and are most intense where genetically engineered Bt cotton has been sold. These are seeds of suicide and seeds of slavery. There are no suicides where farmers use heritage seeds and their own traditional varieties.

Like Gandhi's salt satyagraha, we have undertaken the seed satyagraha – a commitment to not cooperate with patent laws and seed laws which prevent farmers from saving and exchanging seed. Seed freedom is our birthright; without seed freedom there is no food freedom.

Privatisation of the earth's resources – of water, of biodiversity – is the ultimate social and ecological violation of human rights. The earth yields resources to be shared, conserved and used sustainably. The very idea of owning life through patents, of owning and selling water through concessions and commodification, is symptomatic of the deep regression of the human species. Over the years, resisting the enclosure of the commons and aiding their recovery has defined my thinking and my actions in my books, *Biopiracy; Water Wars;* and *Stolen Harvest.*

This is why I have fought against the biopiracy patents on neem, basmati and wheat, and also why I have fought against the commodification of the Ganges and the privatisation of Delhi's water supply. Defending our fundamental freedoms means fighting 'free trade' to protect our seed freedom (bija swaraj); food freedom (anna swaraj); water freedom (jal swaraj); land freedom (bhu swaraj); forest freedom (vana swaraj); and reinventing democracy as Earth Democracy, the democracy of all life as well as the democracy of everyday life.

These new movements for freedom need new learning, new empowerment, new hope. From the seed I learnt lessons of self-organisation and renewal, diversity and democracy; from quantum theory I learnt about non-separability and non-locality; indeterminism and uncertainty; complementarity and non-exclusion; potential and probability. Both the quanta and the seed take us beyond the mechanistic, fragmented, divided, inert, linear, deterministic world of reductionist science, and the industrialisation and commodification of life which is destroying the fragile fabric of the planet and society. In different ways, both the seed and the quanta create a world of relationships, of connectedness, of dynamic evolution and ever-new potential. With the seed we can re-weave the web of life in partnership with other species while increasing their potential to meet basic human needs in sustainable and equitable ways. In the freedom of the seed lies the hope and potential for a better world; and in each of us lie the seeds of our deepest and highest humanity, which comes from returning to our membership in the earth family.

Notes

[1] Renowned social activist, freedom fighter and Gandhian from Tehri district in Uttarakhand.

[2] A European disciple of Gandhi, she was known as Madeleine Slade before she left her home in England in the 1920s to live and work with Mahatma Gandhi. She devoted her life to human development and the advancement of Gandhi's principles.

[3] Born Catherine Mary Heilman, Sarala Behn, another European disciple of Gandhi, was a social reformer whose work in the Kumaon region of Uttarakhand created awareness about the environmental destruction in the Himalayan forests of the state.

[4] Noted Gandhian and eco-activist, Sunderlal Bahuguna is known for leading Chipko in the 1970s and the anti-Tehri dam movement in the 1990s.

[5] Social and environmental activist and wife of Sunderlal Bahuguna, Bimla Bahuguna, also known as Bimla Behn, was among those instrumental in making Chipko a women-led movement.

[6] Gandhian environmentalist, leader of the Chipko movement and founder of Dashauli Swarajya Seva Sangh (DGSS). He is a recipient of the Padma Bhushan, the Magsaysay Award and the Gandhi Peace Prize.

[7] Grassroots community leader from Uttarakhand's Chamoli district who played an important role in the Chipko movement.

[8] A teacher, activist and environmentalist, Negi has been a prominent figure in several social and eco-movements in the Tehri region. He took active part in the Chipko movement, was at the forefront of the agitation against the Tehri dam and mobilised villagers for anti-liquor and anti-quarrying protests. He set up Yuvak Sangh in Tehri to positively channelise the local youth.

Trees of Life
Saving the
Forest

BEFORE LEAVING FOR CANADA TO DO MY PHD ON THE foundations of quantum theory, and while waiting for a bus to get to Delhi, I started talking to the *chaiwala* at the dhaba in Dehradun about my sadness and pain at seeing the oak forests disappear. But he said, 'Now there is hope. Chipko has started,' and he told me about the women who were stopping logging in Reni and other places. I had to leave and catch my flight to Canada, but I pledged to return and find Chipko.

The first time I returned home, we went to the Navjivan Silyara Ashram in Garhwal, which had been started by Sunderlal Bahuguna and his wife, Bimla Bahuguna. From Delhi we took the bus to Tehri, on to Dhanolti, then to Silyara up the Bal Ganga valley, 35 miles from Tehri. Then we walked up a steep forest path to the ashram. Sunderlalji and Bimladi had named the ashram 'Navjivan' (new life), meaning a resurgence and regeneration of living through non-violence, compassion and cooperation.

I spent every vacation as a volunteer with Chipko, going on padyatras (walking tours), visiting communities that were resisting logging. We worked with activists like Dhoom Singh Negi, who was a school teacher. I met the amazing, strong, resilient women of Garhwal, who became my teachers in biodiversity and ecology. From Chipko emerged the slogan that forests are not timber mines but sources of soil, water, and oxygen, which are the basis of life; and from Chipko emerged the call 'Social Forestry, Not Commercial Forestry'.

When I returned to India and joined the Indian Institute of Science in Bangalore to work with Prof G.N. Ramachandran,[1] I observed a strange phenomenon. For some reason farm lands were being planted with eucalyptus trees, and I wanted to get to the bottom of this. The Institute supported our research into this practice, which we carried out during 1980-1981. I often joke that I found the World Bank hiding behind the eucalyptus tree; the Bank had usurped the Chipko slogan of 'Social Forestry' and was using it to give loans to farmers to plant eucalyptus, to be used as raw material for the paper and pulp industry. The entire eucalyptus crop in Karnataka was intended for one factory – Harihar Polyfibre.

We published our research as a report, 'The Ecological Audit of Eucalyptus Plantations', which was widely circulated and was noticed, among others, by Mohd. Idris, the founder of Third World Network (TWN), Consumers Association of Penang (CAP), and Sahabat Alam Malaysia (SAM). SAM was actively resisting the destruction of rainforests in Sarawak, for which SAM-Sarawak received the Right Livelihood Award in 1988 'for their exemplary struggle to save the tropical forests of Sarawak'.

In the mid-1980s, Mohd. Idris invited me to Penang for a meeting dedicated to protecting our tropical forests, because the World Bank had taken its flawed social forestry project and elevated it to an $8 billion Tropical Forestry Action Plan. The TWN asked me to write a book for the movement, which was published as *Forestry Crisis and Forestry Myths: A Critical Review of Tropical Forests: A Call for Action*.

That historical Penang gathering led to the creation of the World Rainforest Movement with its secretariat at the

TWN. The Rainforest Action Network, led by Randy Hayes, was also born of the World Rainforest Movement and the Penang meeting. Penang emerged as the hub of major movements from the mid-1980s to the mid-1990s. The global movement for indigenous approaches to science and technology evolved into the Multiversity. When I started Bija Vidyapeeth/Earth University at Navdanya, I invited Mohd. Idris to inaugurate it, along with leading ecologist Edward (Teddy) Goldsmith; Satish Kumar, activist and founder of Schumacher College; and Sulak Sivaraksha, the legendary Buddhist monk from Thailand who received the Alternative Nobel Prize in 1995 and set up a network of engaged Buddhists. The movement against the GATT/WTO drew strength from our meetings in Penang.

Teddy Goldsmith published our eucalyptus report in *The Ecologist*, the ecological magazine he founded in 1970. In 1972, coinciding with the first environment conference in Stockholm, he published *A Blueprint for Survival* which sold over 750,000 copies. In 1973, he launched the Green Party, initially known as the People Party; we could call him the founder of the contemporary ecology movement, and his inspiration came from India and Gandhi. His book, *The Way*, remains the best distillation of our ecological philosophy of Rta, the path of Right Livelihood.

Mohd. Idris, a proud Tamil Muslim, had migrated from Tamil Nadu to Malaysia. He was a strong resister against colonialism of every hue, including the colonisation of our minds, our science and technology, our knowledge, and our cultures. He wore a lungi and kurta and chappals (slippers) on his feet. I remember, once, when we invited him for a lecture tour to India as part of our mobilisation against GATT/WTO (which we considered another strategy for recolonisation), one of the lectures was in the Calcutta Club which was still governed by colonial rules that forbade entry in lungi and chappals. Mohd. Idris refused to change his clothes and wear trousers, so the Calcutta Club was the loser.

He became 'Uncle Idris' for our generation and inspired the best young people in Malaysia who had studied in Cambridge and Oxford. Martin Khor, an economist, led TWN before he was appointed Executive Director of the South Centre in Geneva. Meena Raman, a lawyer, later married Martin, and she continues to head SAM. Chee Yoke Ling, also a lawyer, took over TWN when Martin moved to Geneva.

On May 17, 2019, Uncle Idris passed on. On April 1, 2020, Martin Khor lost his struggle against cancer.

*****

By 1985, the movement to protect forests had grown considerably. Friends of the Earth (UK), headed by Jonathon Porritt, invited José Lutzenberger (who later became Environment Minster of Brazil), Wangari Maathai (Kenyan activist and the first African woman to win the Nobel Peace Prize), and me to London for a big public meeting to draw attention to the crisis in forestry. The same year, Wangari and I introduced the connection between women and the environment at the UN End of Decade Conference on Women in Nairobi. My first book, *Staying Alive*, published by the feminist press Kali for Women, was a direct consequence of that meeting in Nairobi.

In 1986, the forest movement led by *The Ecologist* collected 3,000,000 signatures demanding an emergency meeting of the United Nations Security Council on global deforestation. Teddy Goldsmith and 20 campaigners took the signatures in wheelbarrows to the UN in New York, which was followed two days later by a meeting with a group of US senators headed by Al Gore.

Over the years, Teddy became a dear, dear friend. He was brilliant, full of joy and jokes, but utterly serious about his commitment to protect the earth. In 1986, he was with me in Rome for a conference hosted by the Society for International Development (SID). Having written the critique of the World Bank forestry projects, I was invited to a debate with Barber Conable, President of the World Bank. Teddy had published a special issue of *The Ecologist* that featured the five most destructive projects of the World Bank; it was this issue that defined the World Bank as an ecologically destructive force on the planet. Teddy was standing by the door of the conference hall with his suitcase filled with copies of *The Ecologist*. As Conable entered, he caught him by the tie, shoved *The Ecologist* in his face, and said, 'You are a murderer!' Today, the terms ecocide and genocide, linked to the destruction of nature, have entered the vocabularies of movements for the protection of nature and of people's rights.

*****

Aranyani: the forest as the feminine principle

Forests have always been central to Indian civilisation. They have been worshipped as Aranyani, the Goddess of the

Forest, the primary source of life and fertility, and the forest as a community has been viewed as a model for societal and civilisational evolution. The diversity, harmony and self-sustaining nature of the forest formed the organisational principles guiding Indian civilisation; the *aranya samskriti* (roughly translatable as 'the culture of the forest' or 'forest culture') was not a condition of primitiveness, but one of conscious choice. According to Rabindranath Tagore, the distinctiveness of Indian culture consists of its having defined life in the forest as the highest form of cultural evolution. In 'Tapovan', he writes:

> Contemporary western civilisation is built of brick and wood. It is rooted in the city. But Indian civilisation has been distinctive in locating its source of regeneration, material and intellectual, in the forest, not the city. India's best ideas have come where man was in communion with trees and rivers and lakes, away from the crowds. The peace of the forest has helped the intellectual evolution of man. The culture of the forest has fuelled the culture of Indian society. The culture that has arisen from the forest has been influenced by the diverse processes of renewal of life, which are always at play in the forest, varying from species to species, from season to season, in sight and sound and smell. The unifying principle of life in diversity, of democratic pluralism, thus became the principle of Indian civilisation.

The forest thus nurtured an ecological civilisation in the most fundamental sense of harmony with nature. Such knowledge that came from participation in the life of the forest was the substance not just of the *Aranyakas* or the ancient forest texts, but also of the everyday beliefs of tribal and peasant society. Sacred forests and sacred groves were created and maintained throughout India as a cultural response for their protection. As G.B. Pant reports for the Himalaya in *The Forest Problem in Kumaon*:

> A natural system of conservancy was in vogue; almost every hill-top is dedicated to some local deity, and the trees on or about the spot are regarded with great respect so that nobody dare touch them. There is also a general impression among the people that everyone cutting a tree should plant another in its place.

All religions and cultures of the South Asian region have been rooted in the forests, not through fear and ignorance, but through ecological insight. India's people have traditionally recognised the dependence of human survival on the existence of forests. A systematic knowledge about plants and forest ecosystems was thus generated and informal principles of forest management formulated. It has often been stated that 'scientific' forestry and the 'scientific' management of forest resources in India began with the British. The historical justification for such a statement becomes possible only if one accepts that modern western patriarchal science is the only valid science. In ancient Indian traditions, scientific knowledge of the plant kingdom is evident from such terms as *vriks ayurveda*, which means the science of the treatment of plant diseases, and *vanaspati vidya* or plant sciences. Being derived from the living forest, indigenous forestry science did not perceive trees as just wood; they were looked at from a multifunctional point of view, with a focus on diversity of form and function.

A distinction has also been made between natural and cultivated forests, suggesting that afforestation and regeneration through the planting of trees has always been significant in the renewal of the forest wealth of the region.

Ethnobotanical work among India's many diverse tribes is uncovering the deep, systematic knowledge of forests among them. The diversity of forest foods used in the country emerges from this knowledge. In south India, a study conducted among the Soliga tribe in the Belirangan hills of Karnataka shows that they use 27 different varieties of leafy vegetables at different times of the year, and a variety of tubers, leaves, fruits and roots are used for their medicinal properties by them. A young illiterate Irula boy from a settlement near Kotagiri could identify 37 different varieties of plants, with their Irula names and their different uses.

In non-tribal areas, too, forests provide food and livelihood through critical inputs to agriculture, through soil and water conservation, and through inputs of fodder and organic fertiliser. Indigenous silvicultural practices are based on a sustainable and renewable maximisation of all the diverse forms and functions of forests and trees. This common silvicultural knowledge is passed on from generation to generation, through participation in the processes of forest renewal and of drawing sustenance from the forest ecosystem.

In both forest and agriculture-based economies, it is primarily women who use and manage the produce of forests and trees. In the Himalaya, where tree fodder is predominant in the agricultural economy even today, older women train the younger ones in the art of lopping (pollarding) and of collecting forest produce. In other regions, lopping cycles and practices had evolved to maximise fodder production. Since food gathering and fodder collection has primarily been women's work, women as foragers were critical in managing and renewing the diversity of the forest. Their work was complementary to that of men. The public and common domain of the forest was not closed to women – it was central to supporting life in the 'private' domain, the home and community.

Since it is women's work that protects and conserves nature's life in forestry and in agriculture and, through such conservation work, sustains human life by ensuring the provision of food and water, the destruction of the integrity of forest ecosystems is most vividly and concretely experienced by peasant women. For them, forestry is married to food production; it is essential for providing stable, perennial supplies of water for drinking and for irrigation, and for providing the fertility directly as green manure or as organic matter cycled through farm animals. Women's agricultural work in regions like the Himalaya is largely work in and with the forest, yet it is discounted both in forestry and in agriculture. The only forestry-related work that is accounted for in census data is lumbering mid tree-felling; cutting trees then becomes a source of *roti* or food for the men engaged in lumbering operations. For the women, however, forests are food not in death, but in life. When, for example, women lop trees, they enhance the productivity of the oak forest under stable conditions and under common ownership and control. While an unlopped tree has leaves that are too hard for cattle, lopping makes them soft and palatable, especially in early spring. Maintaining the diversity of living resources is critical to the feminine use of the forest: thus oak-leaf, along with a mixture of dried grasses and agricultural byproducts, is fed to cattle through the late autumn, winter and spring. In the monsoon, the green grass becomes the dominant fodder, and in October and November, agricultural waste such as rice straw, *mandua* straw and *jangora* straw become the primary supply of fodder. Groups of women, young and old, go together to lop for fodder, and expertise develops by participation and through learning-by-doing. These informal forestry colleges of the women are small and decentred, creating and transferring knowledge about how to maintain the life of living resources. By contrast, the visible forestry colleges are centralised and alienating: they specialise in a forestry of destruction, on how to transform a living resource into a commodity and subsequently, cash.

Beginning in the early 1970s in the Garhwal region of Uttar Pradesh, the methodology and philosophy of the Chipko movement that I mentioned earlier quickly spread to Himachal Pradesh in the north, to Karnataka in the south, to Rajasthan in the west, to Orissa in the east and to the Central Indian highlands. Women's environmental action in India preceded the UN's Decade for Women as well as the 1972 Stockholm Environment Conference. Three hundred years ago, more than 300 members of the Bishnoi community in Rajasthan, led by a woman called Amrita Devi, sacrificed their lives to save their sacred *khejri* trees by clinging to them. With that event begins the recorded history of Chipko.

The Chipko movement was popularly referred to as a women's movement, but it is only some male Chipko activists who have been projected into visibility. The contribution of women has been neglected and remains invisible, in spite of the fact that the history of Chipko is a history of the visions and actions of exceptionally courageous women. Environmental movements like Chipko have become historical landmarks because they have been fuelled by the ecological insights and political and moral strengths of women. I will dwell at some length on some of these exceptional women because I have personally been inspired by my interaction with them, and because I feel that it is unjust that the real pillars of the movement are still largely unknown. The experience of these powerful women also needs to be shared to remind us that we are not alone, and that we do not take the first steps; others have walked on this path before us.

In the history of social and political movements, their evolution is generally neglected, and only the end result focussed on. This creates two problems: first, future organisational work does not benefit from the lessons of perseverance and patience born of years of movement-building; people start looking for instant solutions because it is the instant successes that have been sold through pseudo-history. Second, while the historical evolution of movements involves significant contributions from thousands of participants, over extended periods, their climaxes are localised in space and time. This facilitates the appropriation of the movement by an individual or group, who then erases the contributions of others. Movements are major social and political processes, however, and they transcend individual actors. They are significant precisely because they involve a multiplicity of people and events which contribute to a reinforcement of social change.

The Chipko process, as a resurgence of woman power
and ecological concern in the Garhwal Himalaya, is a similar
mosaic of many events and multiple actors. The significant
catalysers of the transformations which made the Chipko
resistance possible have been women like Mira Behn, Sarala
Behn, Bimla Behn, Hima Devi, Goura Devi, Gunga Devi,
Bachhni Devi, Itwari Devi, Chamun Devi and many others.
The men of the movement, like Sunderlal Bahuguna, Chandi
Prasad Bhatt, Ghanshyam Raturi and Dhoom Singh Negi
have been their students and followers. Mira Behn, one
of Gandhi's closest disciples, had moved to the Himalayan
region in the late '40s, and started a cattle centre called
Pashulok between Rishikesh and Haridwar because cattle
are central to sustainable agriculture. Writing to Mira Behn
fifteen days before his death, Gandhi said:

> I see that you are destined for serving the cow and nothing
> else. But I seem to see a vital defect in you. You are unable
> to cling to anything finally. You are a gypsy, never happy
> unless you are wandering.

As Gandhi had expected, Mira Behn moved on from the
ecology of the cow to the ecology of forests and water, to
the links between deforestation and water crises. As she
recollected later,

> Pashulok being situated as it is at the foot of mountains, just
> where the Ganga emerges from the Himalayan valleys, I
> became very realistically aware of the terrible floods which
> pour down from the Ganga catchment area, and I had taken
> care to have all the buildings constructed above the flood
> high-mark. Within a year or two I witnessed a shocking
> flood: as the swirling waters increased, (there) came first

bushes and boughs and great logs of wood, then in the turmoil of more and more water came whole trees, cattle of all sizes, and from time to time, a human being clinging to the remnants of his hut.... The sight of these disastrous floods led me each summer to investigate the area north of Pashulok whence they came. Merciless deforestation as well as cultivation of profitable pines in place of broad-leaf trees was clearly the cause.

During her stay in Garhwal, Mira studied the environment intimately and derived knowledge about it from the local people. From the older ones she learnt that, earlier, Tehri Garhwal forests consisted largely of oak, and Garhwali folksongs, which encapsulate collective experience and wisdom, tell repeatedly of species such as *banj* and *kharik*. In Mira's view, the primary reason for degeneration in this region was the disappearance of the *banj* trees. According to her, if the catchment of the Ganga was not once again clothed with *banj*, floods and drought would continue to get aggravated. The issue was not merely one of planting trees, but of planting *ecologically appropriate* trees.

Mira Behn's ecological insights were inherited by Sunderlal Bahuguna, who had worked with her in the Bhilangana valley. Inspired by Sri Dev Suman, he had joined the independence struggle at the tender age of 13, and was Congress Secretary of Uttar Pradesh at the time of Independence. In 1954 he married Bimla Behn, who had spent eight years with Sarala Behn. Sarala Behn had started an ashram for the education of hill women in Kausani, and her full-time commitment was to make them recognise that they were not beasts of burden but goddesses of wealth, since

they rear cattle and produce food, performing 98 per cent
of all labour in farming and animal husbandry. Influenced
by Sarala Behn's ideas of women's freedom, Bimla agreed
to marry Bahuguna only if he left the Congress Party and
settled down in a remote village so that they could awaken
the hill people by living with and through them. Twelve years
after the establishment of their Silyara Ashram, Sunderlal and
Bimla Bahuguna wrote: 'One of us, Sunderlal, was inspired
to settle in a village by Mira Behn and the other, Bimla, was
inspired by living continuously with Sarala Behn.' Bahuguna,
in turn, drew in others like Raturi, Bhatt and Negi to lend
support to the Chipko movement. As he often said, 'We are
the runners and messengers – the real leaders are the women.'
While the philosophical and conceptual articulation of the
ecological view of the Himalayan forests has been done by
Mira Behn and Bahuguna, the organisational foundation
for it being a women's movement was laid down by Sarala
Behn, with Bimla Behn in Garhwal and Radha Bhatt[2] in
Kumaon. Sarala Behn knew that the ethics of sharing, of
producing and maintaining life, which women conserved in
their activity, was the countervailing force to the masculinist
morality of the market which came as 'development' and
created a cash economy but also created destitution and
drunkenness. The early women's movement in Uttarakhand
was therefore an anti-alcohol movement aimed at controlling
alcohol addiction among men, who earned cash incomes
from felling trees with one hand and lost the cash to liquor
with the other. For the women, drunkenness meant violence
and hunger for their children and themselves, and it was

the organisational base created among them through the anti-alcohol movement that was inherited by Chipko.

I have written at length about the Chipko movement and about how women's involvement led to its radical shift in focus: from local *control* over forest resources to *protecting* the forest itself, for survival. It was this articulation that impelled the women to hug the trees, cling to them and save them from the axe-men.

There are two paradigms of forestry in India today – one life-enhancing, the other life-destroying. The life-enhancing paradigm emerges from the forest and the feminine principle; the life-destroying one from the factory and the market. The former creates a sustainable, renewable forest system, supporting and renewing food and water sources. *The maintenance of conditions for renewability is the primary management objective of the former*, while the maximising of profits through commercial extraction is the primary management objective of the latter. Since the maximising of profits is consequent upon the destruction of the conditions of renewability, the two paradigms are cognitively and ecologically incommensurate.

It is these two distinct knowledge and economic systems which clashed in 1977 in the Adwani forests when the Chipko movement became explicitly an ecological *and* feminist movement. The women, of course, had always been the backbone of Chipko and for them the struggle was ever the struggle for the living, natural forest. But in the early days when the action was directed against removing the non-local forest contractors, local commercial interest

had also been part of the resistance. However, even after
the non-local private contractors were removed and a
government agency (the Uttar Pradesh Forest Corporation)
started working through local labour contractors and forest
co-operatives, *the women continued to struggle against the
exploitation of the forests*. It did not matter to them whether
the forest was destroyed by outsiders or their own men.
The most dramatic turn in this new confrontation took
place when Bachhni Devi of Adwani stood against her own
husband, who had obtained a local contract to fell the forest.
The forest officials arrived to browbeat and intimidate the
women and the Chipko activists, but found the women
holding up lit lanterns in broad daylight. Puzzled, the forester
asked them about their intention. The women replied, 'We
have come to teach you forestry.' He retorted, 'You foolish
women, how can you who prevent felling know the value
of the forest? Do you know what forests bear? They produce
profit and resin and timber.' And the women immediately
sang back in chorus:

> *What do the forests bear?*
> *Soil, water and pure air.*
> *Soil, water and pure air*
> *Sustain the earth and all she bears.*

The main thrust of conservation struggles like Chipko
is that forests and trees are life-support systems, and should
be protected and regenerated for their biospheric functions.
The crisis mind, however, sees the forest and trees as weed,
or to be valued commercially, and converts even afforestation
into deforestation and desertification. From life-support

systems, trees are converted into green gold – all planting is motivated by the slogan 'Money grows on trees.' Whether it is schemes like social forestry or wasteland development, afforestation programmes are conceived at the international level by 'experts' whose philosophy of tree-planting falls within the reductionist paradigm of producing wood for the market, not biomass for maintaining ecological cycles or satisfying local needs of food, fodder and fertiliser. All official programmes of afforestation, based on heavy funding and centralised decision-making, act in two ways against the feminine principle in forestry – they destroy the forest as a diverse and self-reproducing system, and devastate it as commons, shared by a diversity of social groups with the smallest having rights, access and entitlements.

'Social' forestry and the 'miracle' tree

Social forestry projects are a good example of single-species, single commodity production plantations, based on reductionist models which divorce forestry from agriculture and water management, as well as needs from markets.

A case study of the World Bank–sponsored social forestry in Kolar district of Karnataka is an illustration of reductionism and maldevelopment in forestry being extended to farmland. Decentred agro-forestry, based on multiple species and private and common tree stands, has been India's age-old strategy for maintaining farm productivity in arid and semi-arid zones. The *honge*, tamarind, jackfruit and mango, the *jola, gobli, kagle* and bamboo traditionally provided food and fodder, fertiliser and pesticide, fuel and small timber. The backyard of each rural home was a nursery, and each peasant woman the silviculturist. This invisible, decentred agro-forestry model was significant because the humblest of species and the smallest of people could participate in it, and with space for the small, *everyone* was involved in protecting and planting.

The reductionist mind took over tree-planting with 'social forestry'. Plans were made in national and international capitals by people who could not know the purpose of the *honge* and the neem, and instead saw them as weeds. The experts decided that indigenous knowledge was worthless and 'unscientific', and proceeded to destroy the diversity of indigenous species by replacing them with row after row of eucalyptus seedlings in polythene bags in government nurseries. Nature's locally available seeds were laid waste; people's locally available knowledge and energies were laid waste. With imported seeds and expertise came the import of loans and debt, and the export of wood, soils – and people. Trees, as a living resource that maintain the life of the soil and water and of local people, were replaced by trees whose dead wood went straight to a pulp factory hundreds of miles away. The smallest farm became a supplier of raw material to industry and ceased to be a provider of food to local people.

Women's work, linking the trees to the crops, disappeared and was replaced by the work of brokers and middlemen who bought the eucalyptus trees on behalf of industry. Industrialists, foresters and bureaucrats loved the eucalyptus because it grows straight and is excellent pulp-wood, unlike the *honge*, which shelters the soil with its profuse branches and dense canopy, and whose real worth is as a living tree on a farm. The *honge* could be nature's idea of the perfect tree for arid Karnataka. It has rapid growth of precisely those parts of the tree, the leaves and small branches, which go back to the earth, enriching and protecting it, conserving its moisture and fertility. The eucalyptus, on the other hand, when perceived ecologically, is unproductive, even negative, because this perception assesses the 'growth' and 'productivity' of trees in relation to the water cycle and its conservation; in relation to soil fertility; and in relation to human needs for food and food production. The eucalyptus has destroyed the water cycle in arid regions due to its high water demand and its failure to produce humus, which is nature's mechanism for conserving water. Most indigenous species have a much higher biological productivity than the eucalyptus when one considers water yields and water conservation. The non-woody biomass of trees has never been assessed by forest measurements and quantification within the reductionist paradigm, yet it is this very biomass that functions in conserving water and building soils. It is little wonder that Garhwal women call a tree *dali* or branch, because they see the productivity of the tree in terms of its non-woody biomass, which functions critically in hydrological and nutrient cycles within the forest, and through green fertiliser and fodder in cropland.

Susan Griffin, in *Woman and Nature*, parodied the reductionist mind when she wrote:

> The trees in the forest should be tall and free from knot-causing limbs for most of their height....They should be straight.... Trees growing in the forest should be useful trees. For each tree ask if it is worth the space it grows in. Aspen, scrub pine, chokeberry, black gum, scrub oak,

dogwood, hemlock, beech are weed trees which should be eliminated....

For harvesting trees, it is desirable that a stand be all of the same variety and age. Nothing should grow on the forest floor, not seedling trees, not grass not shrubbery.

She contrasts this uniformity with the logic of diversity in the forest as feminine. The voices of women join the voices of nature:

> The way we stand, you can see we have grown up this way together, out of the same soil, with the same rains, leaning in the same way toward the sun.... And we are various, and amazing in our variety, and our differences multiply, so that edge after edge of the endlessness of possibility is exposed. You know we have grown this way for years. And to no purpose you can understand. Yet what you fail to know we know, and the knowing is in us, how we have grown this way, why these years were not one of them heedless, why we are shaped the way we are, not all straight to your purpose, but to ours. And how we are each purpose, how each cell, how light and soil are in us, how we are in the soil, how we are in the air, how we are both infinitesimal and great and how we are infinitely without any purpose you can see, in the way we stand ... each moment heeded in this cycle, no detail unlovely.★

It is such a recovery of life in diversity, of a diversity shared and protected, that the invisible Chipko activist struggled for. Giving value and significance to Prakriti, to nature as the source, to the smallest element of nature in

★ S. Griffin, *Women and Nature*, London: The Women's Press, 1984.

its renewal, giving value to collective needs, not private action, the village women of Kangod, Sevalgoan and Rawatgaon worked in partnership with nature to recreate and regenerate. Without signboards, without World Bank loans, without wire-fencing, they worked to allow nature's play in reproducing the life of the forest – grasses and shrubs, small trees and big, each useful to nature if not to man, all coming alive again.

The annihilation of this diversity has destroyed women's control over the conditions of producing sustenance. The many colonisations – through 'reserved' forests, through 'social forestry', through 'wasteland' development – have implied not forest development but the maldevelopment of both forestry and agriculture. A maldeveloped forestry has meant new resources and raw material supplies for industry and commerce; for nature and women it has meant a new impoverishment, a destruction of the diverse means of production which both provide sustenance in food and water and reproduce society. The Chipko struggle was a struggle to recover the hidden and invisible productivity of vital resources, and the invisible productivity of women to recover their entitlements and rights, to have and provide nourishment for sustained survival, and to create ecological insights and political spaces that do not destroy fundamental rights to survival. The Chipko women provided a nonviolent alternative in forestry to the violence of reductionist forestry with its inherent logic of dispensability, and took significant steps towards recovering their status as the *other* silviculturists and forest managers, who participate in nature's processes instead of working against them, and share nature's wealth for basic needs instead of privatising it for profit.

*****

My critique of the World Bank's Tropical Forestry Action Plan was based on the work done on the ground by diverse ecological movements in India and Southeast Asia, as well as on the experience of similar projects and action plans in Africa.

The crisis in tropical forest resources is recognised as the world's most severe ecological crisis, having a long-term impact on the potential for economic development. This is so because the food security of the entire world is linked to the genetic resources of tropical forests, and the economic well-being of the majority of the world's people who live in what used to be called the Third World is linked to their ecological stability. The reckless exploitation of tropical forests, among other natural resources, provided the material basis for industrialisation and economic growth in the colonial and post-colonial periods. Today, the cumulative impact of this reckless over-exploitation has created a critical and nearly irreversible condition of ecological degradation. The famines in Africa and other arid regions have turned the world's attention to the high ecological and social costs of tropical deforestation. It has become the central concern for governments, development agencies and ecology movements all over the world. Yet, the focus on tropical deforestation and the resulting reversal of the process does not automatically translate into the protection of tropical forests and those who depend on them for survival. As long as misconceptions about the nature of tropical forest ecosystems prevail, and as long as the causes of tropical deforestation are wrongly located, the degradation cannot be arrested. As a result, tropical ecosystems will continue to be threatened.

Normally, forests are identified only with the economy associated with commercial industrial use; but the crisis of tropical forestry needs to be understood in the light of the fundamental economies associated with them – nature's economy of essential ecological processes and the market economy of industrial-commercial demands.

Nature's economy generates a demand on the products of the forest in terms of maintaining the stability of the soil systems and the hydrological balance of forest ecosystems. For example, the production of humus is essential for conserving soil and water. In ecologically sensitive ecosystems like upland watersheds, this economy assumes the most importance and should get the necessary priority in forest management. Neglecting this entails a huge amount of negative externalities to the national exchequer as relief for regular floods and drought (which are easily described as nature's fury), pushing the ecological basis of such developments into the background.

The survival economy of basic needs satisfaction reflects the requirement of forest biomass for people living inside and in the vicinity of the forests, in terms of fuelwood, fodder, fruits, nuts, green manure, small timber, etc. In forest areas where human settlements have existed or exist in the vicinity, the requirement of the survival economy has been satisfied all along without major ecological damage. In regions where forests no longer exist and land has been under the plough for centuries, farm trees and agro-forestry systems have provided forest inputs to the farm economy. Under certain situations, the pressure of the survival economy can be substantial and its neglect in development policies can lead to an unexpected and rapid degradation of forest resources.

The market economy of industrial-commercial demands constitutes the forest biomass demand of the total market system in the formal market economy. It includes the demand for pulpwood, plywood, furniture, construction timber, etc., as well as fuelwood for urban consumers.

There are four pervasive myths backing international forestry programmes created by international 'aid', which militate against their becoming strategies for the ecological and economic recovery of marginalised communities. They are:

(*i*) people, not profits, are the primary cause of tropical deforestation;

(*ii*) the 'developed' world has protected its forests and must teach conservation to countries in the tropics;

(*iii*) commercial forestry based on privatisation can solve the scarcity problems of the poor; and

(*iv*) commercial afforestation can guarantee ecological recovery.

These myths were revived in the World Resources Institute (WRI) report 'Tropical Forests: A Call for Action', published for the World Bank and the United Nations Development Programme (UNDP) in 1985. We, in the Global South in general and in India in particular, are familiar with these myths. They were the political tools used by the British for the colonisation of common forest resources. The centres of exploitation and planned destruction might have shifted from the East India Company and the Crown in London a century ago to the World Bank in Washington in contemporary times, but the logic of colonisation has not changed. The British, too, talked of 'forest conservation' while creating a policy for deforestation. The World Bank, following the same pattern, talks of 'conserving' tropical ecosystems while financing projects that will destroy them.

The World Bank's call for action for tropical forests was inconsistent with the socio-ecological imperative of sustainability and survival in the tropics. It threatened to create new forms of poverty for the poor, and new forms of ecological destabilisation in the tropics, even while the Bank's forestry projects were legitimised on environmental grounds and in the cause of poverty alleviation. But legitimising and packaging are not the same as content, and all its forestry programmes, all its tropical forestry action plans, were created in the Bank's vision of the theology of the market in which neither the poor nor nature have a role – except as victims.

The WRI report revived the myth that it is local people who destroy tropical forests. It states,

> ... it is the rural poor themselves who are the primary agents of destruction as they clear forests for agricultural land, fuelwood, and other necessities. Lacking other means to meet their daily survival needs, rural people are forced to steadily erode the capacity of the natural environment to support them.

The reality of deforestation, however, is quite different. The tribals of Bastar in central India, for example, had protected their forests over centuries; but a World Bank project for 'forest development' became a major cause of deforestation in the region. The project came to India as the Bank's first intervention in forestry, the Madhya Pradesh Forestry Technical Assistance Project (Credit 608-IN, December 1975) and was directed primarily towards the development of plantations for the pulp and paper industry.

The Bastar project was part of a trend to convert natural forests into commercial plantations so that the biomass produced could no longer benefit the original forest-dwellers. The tribal sustenance base in cane and bamboo for basket-weaving along with mangoes, tamarind, jackfruit, *mahua* and edible berries are all destroyed when natural forests are replaced by monoculture plantations of eucalyptus or tropical pine. The WRI report cited the worldwide distribution of tropical pine as a 'success', an example of scientific achievements in forestry research by the Commonwealth Institute. The Bastar Tropical Pine Project was planned at Rs 96,000,000 to convert 8,000 ha of natural forests in the Bastar hills into pine plantations to feed the paper and pulp industry. It was finally shelved due to serious resistance, because for local tribals this was an example of a forestry disaster, not of success. It was based not on scientific knowledge but on ecological ignorance of the forest ecosystem and of the tribals' integration with that ecosystem. It was the ultimate erosion of the tribal right to forests as survival systems, a project aimed at changing the very character of the forests.

The new strategy of blaming the people for environmental destruction in general, and deforestation in particular, arose from agencies like the WRI, the World Bank and the Food and Agriculture Organization (FAO) to facilitate the transfer of resource control and forest management from local people to the state, from the countries of the South to pseudo-experts in international aid agencies in the North. As Lloyd Timberlake[3] has pointed out, Africa is not dying because Africans are ignorant, 'Africa is dying because in its ill-planned, ill-advised attempt to "modernise" itself it has cut itself into pieces.'

Under the Third Five Year Plan (1968-73), Ethiopia, for example, spent only one per cent of its total expenditure on peasant agriculture, emphasising instead the rapid development of large-scale commercial farms producing crops for export. Tractors, pesticides, fertilisers were exempted from import duty. Multinationals making agrarian investments of $200,000 or more were given a three-to-five-year income tax holiday. Commercial development of the Awash Valley was part of the plan. By 1970, 60 per cent of the land brought under cultivation in the Awash Valley had been devoted to cotton production, while sugar plantations claimed another 22 per cent of cultivated area. To make way for these multinational-managed commercial farms, the government forcibly evicted Afar pastoralists from their traditional lowland pastures. The Afars were thus pushed into the fragile uplands which were rapidly over-grazed and degraded. The degradation of the Ethiopian highlands needs to be viewed in this context of the introduction of commercial, export-oriented agriculture in the lowlands, and the consequent displacement of nomads and peasants. It is not local ignorance but the global control and exploitation of land and forests that catalysed the tropical forest crisis.

In India, the World Bank financed the destruction of tropical forests through dams on the Narmada river, and through mining and energy generation at Singrauli in central India. In Brazil, it has financed the destruction of the Amazonian rainforests through the Big Carajas project which comprises 10 per cent of the Amazon. The project involves mining and the construction of about 30 large dams like the Tucurui and Balbina. Friends of the Earth (Brazil) have observed that

> 'The Tropical Forests: A Call for Action' plan unfortunately does not go to the roots of the problem ... whereas it makes almost no reference to the economic/political power that rules our civilisation, that plans, decides and puts into practice the projects which demolish whole tracts of Tropical Rain Forests (TRFs), the plan states that 'tragically it is the rural poor themselves who are the major agents of destruction'.

When, after clearing the forest, the soil is exhausted within three years, the small settlers just move further into the forest. But who is to be blamed for this situation? Clearly, it can't be the poor who never had any role in the decision-making.

In Brazil, the plan foresaw investments of US $400 million for fuelwood and agro-forestry; US $325 million for forest management for industrial uses; and US $50 million for the conservation of TRF ecosystems. So the bulk of the money was to be spent on industrial development and to strengthen official departments. What can be the credibility of a government that has already been proved to ignore ecological imperatives, indigenous rights and democratic principles? The World Bank itself had noted its experience in the case of the Polonoroeste, when the Brazilian government failed to accomplish what had been set out in the plan. Its Call for

Action plan declared the following: 'Bring under control and management five million ha of Amazon forest.' What did this mean? Who *knows* how to manage this forest? Who was expected to, and would be able to, control this management of five million ha, and who would be the beneficiaries? Where were the studies underlying this project? Who lives there today?

Till now, the only people who have succeeded in developing an economic activity based on the products of the forest *without* destroying it have been local rubber-tappers and the 'castanheiros', who collect the Castanhado Para. But they are modest people, they 'live lightly on the resources of the Earth', they 'are alien to the greed that marks our present western society'.

The WRI report was an example of Orwellian doublespeak in which ecological destruction is called 'environmental protection'. It was a revival of the myth that the destroyers of tropical forests will protect them, that peoples alienated from and ignorant of tropical forests will prescribe to those who have generations of knowledge about local forest ecosystems. The arrogant ignorance of Washington experts in the statement that 'solutions are known' failed to accept that project after project which originated in Washington was shown to be socio-ecologically flawed. The World Bank definitions of 'productivity' and 'development' have been systematically found to be in direct contradiction to concepts of 'productivity' and 'development' from the perspective of local basic needs and sustainability.

Its 'social forestry' projects in India were often cited as a success story, launched ostensibly to correct the one-dimensional commercial forestry that has made forest resources more scarce for the poor, and to enhance the availability of organic fertiliser, food, fodder and fuel for local populations. The result, however, was a further erosion of rights and resources. More than 90 per cent of tree-planting under social forestry has been of eucalyptus, nearly all of it has been on fertile agricultural land, and all of it has been marketed to urban industrial centres, especially the pulp industry. The erroneous organisational assumption that is at the root of such afforestation projects lies in the *organisational equivalence* of different management structures. It is assumed that the outcome is the same whether a resource is managed collectively or privatised, by MNCs or local tribals. This assumption then enables the growth of organisational structures which lend themselves more easily to serving the interests of the economically more powerful.

However, ecological audit of different afforestation programmes has shown that the organisational structures which are effective tools in a market economy that runs primarily on commercial objectives can actually be inconsistent with the needs of the survival economy. For local needs, multipurpose tree-planting is required for food, fuel, fibre, fodder, fertiliser, oilseeds and medicines. In biomass production, as in any other form of production, the relations of production determine the relations of distribution: how production is organised determines how the produce is distributed. If production is organised in accordance with the logic of the market economy it will get distributed according to that logic; purchasing power, not need, will determine access and entitlement. The poor who have no purchasing power, but whose needs are the greatest, cannot register their demands through the organisational structures of the market. Tree-planting success in the commercial economy does not automatically translate into successful tree-planting for the survival economy. Further, at a historical point in the evolution of markets, when plantations bring a 16 times higher return on investment than food production, tree-planting for 'green gold' can disastrously aggravate the real energy crisis of food scarcity.

In the World Bank's plan for tropical forests there was no organisational or ecological difference between the successful case studies of 'farm and community forestry' in India and industrial forestry in the Philippines. Whereas the Bank categorised the production of industrial pulpwood by farmers as 'social' forestry projects in India, it described similar projects as 'industrial' forestry in the Philippines.

Between May 1974 and 1985, the World Bank provided US $2 million to the Development Bank of the Philippines for supplying 284,000 cubic metres of pulpwood to the Paper Industries Corporation of the Philippines (PICOP), which was formed within the Serrano group of companies for manufacturing pulp and paper. The loan was used to finance farmers shifting from food production to fast-growing pulpwood trees, an eight-year rotation on 80 per cent of their land-holding. The PICOP supplied seedlings at cost in exchange for first rights to mature pulpwood. The World Bank subsidised rich industrialists in the South at the cost of the poor, who lost livelihoods and were further burdened with the social and economic cost of repaying debt with interest on World Bank loans.

The Bank's Tropical Forestry Action Plan was a recommendation for the expansion of these destructive activities. In India, it recommended the expansion of the Bank's 'social' forestry programme for commercial wood production, which desertified farmlands and displaced peasants at the rate of 200 man-days per hectare per year at the cost of US $500 million. At the cost of another US $190 million it recommended the exploitation of 30 million hectares of natural forest and the establishment of industrial plantations at the rate of 240,000 hectares per year. In Brazil, it recommended an investment of US $400 million to convert natural forests into plantations as a 'Fuelwood and Agroforestry' Project and another US $325 million for managing five million hectares of Amazon forests for industrial wood production, and establishing 320,000 hectares of industrial plantations over a five year period. If the Action Plan was operationalised in India and Brazil alone, the people would have borne the burden of US $1,415 million in loans to destroy millions of hectares of natural forests and prime farmlands. This did not include the destruction of watersheds through commercial tree-planting (US $500 million for India).

The systemic inability of forestry projects to meet ecological and social needs arises, in my view, from wrong scientific and organisational assumptions. Scientifically, it is assumed that all trees and all forest ecosystems are ecologically equivalent. This assumption of the *ecological equivalence* of different tree species and silvicultural systems overlooks the difference between the imperatives of temperate zone and tropical zone forest management; between commercial and conservation forestry; as well as differences in the ecology of different tree species. The assumption that temperate zone practices are suitable for tropical ecozones has been identified as a significant cause of the erosion of forest resources even when 'scientifically' managed. It also misses the physiological and architectural diversity in tree species which is tailored to local ecosystem diversity. Trees have their own water relations, nutrient relations, and patterns of partitioning organic matter which are determined by their native habitat. The wrong species in the wrong place can undermine essential ecological processes and basic needs satisfaction. Different species have different ecological and social impacts, and different silvicultural practices have different ecological and social effects. All tree-planting is not rebuilding nature's economy. Some plantations cause major dislocations in nature's processes; to subsume all tree-planting under a uniform category of 'green cover', falsely assuming ecological equivalence, is to ignore the diversity in nature and the diverse human needs that nature's diversity supports.

*****

I want to tell a story about a tree in India, ancient and revered across the country for its many life-affirming and medicinal properties. It is the neem, known in Sanskrit as *Sarva Roga Nivarini*, or the 'curer of all ailments'.

Of all the plants that have proved useful to humanity, a few are distinguished by their astonishing versatility. The coconut palm is one, bamboo another. In the more arid areas of India, this distinction is held by a hardy, fast-growing evergreen of up to 20 metres in height – *Azadirachta indica*, commonly known as the neem tree. The neem's many virtues are largely attributable to its chemical constituents. From its roots to its spreading crown, the tree contains a number of potent compounds, notably an astringent found in its seed named *azadirachtin*. It is this astringency that makes it useful in so many fields.

Neem's bark, leaves, flowers, seeds and fruit pulp are used to treat a wide range of diseases and complaints, ranging from leprosy and diabetes to ulcers, skin disorders and constipation. Its twigs are used by millions of Indians as an antiseptic toothbrush. Its oil is used in the preparation of toothpaste and soap. Neem oil is known to be a potent spermicide and is considered to be 100 per cent effective when applied intra-vaginally before intercourse. Intriguingly, it is also taken internally by ascetics who wish to abate sexual desire.

Besides being hard and fast-growing, neem's chemical resistance to termites makes it a useful construction material. Neem oil is used as lamp oil, while the fruit pulp is useful in the manufacture of methane. Neem cake, the residue from the seed after oil extraction, is fed to livestock and poultry, while its leaves increase soil fertility. Most importantly, neem is a potent insecticide, effective against about 200 insects, including locusts, brown plant-hoppers, nematodes, mosquito larvae, Colorado beetles and boll weevils.

There are some 14 million neem trees in India and the age-old village techniques for extracting the seed oil and pesticidal emulsions do not require expensive equipment. A large number of different medicinal compounds based on neem are commonly available.

In the last 70 years, considerable research has been done on the properties of neem, carried out in institutes ranging from the Indian Agricultural Research Institute and the Malaria Research Centre to The Energy and Resources Institute (TERI) and the Khadi and Village Industries Commission (KVIC). Much of this research was fostered by Gandhian movements, such as the Boycott of Foreign Goods movement, which encouraged the development and manufacture of local Indian products. A number of neem-based commercial products, including pesticides, medicines and cosmetics have come into the market in recent years, some of them produced in the small-scale sector under the banner of the KVIC, others by medium-sized laboratories. However, there has been no attempt to acquire proprietary ownership of the formulae since, under Indian law, agricultural and medicinal products are not patentable.

For centuries, the western world ignored the neem tree and its properties: the practices of Indian peasants and doctors were not deemed worthy of attention by the majority of British, French and Portuguese colonialists. However, in the last few years, growing opposition to chemical products in the West, in particular to pesticides, has led to a sudden enthusiasm for the pharmaceutical properties of neem.

In 1971, US timber importer Robert Larson observed the tree's usefulness in India and began importing neem seeds into his company headquarters in Wisconsin. Over the next decade he conducted safety and performance tests upon a pesticidal neem extract called Margosan-O, and in 1985 received clearance for the product from the US Environmental Protection Agency (USEPA). Three years later he sold the patent for the product to the multinational chemical corporation, W R Grace and Co. Since 1985, over a dozen US patents were taken out by US and Japanese firms on formulae for stable neem-based solutions and emulsions, and even for a neem-based toothpaste. At least four of these were owned by W R Grace, three by another US company, the Native Plant Institute, and two by the Japanese Terumo Corporation.

Having garnered their patents, and with the prospect of a licence from the USEPA, Grace set about manufacturing and commercialising their product by establishing a base in India. The company approached several Indian manufacturers with proposals to buy up their technology or to convince them to stop producing value-added products and instead supply the company with raw material.

In many cases, Grace met with a rebuff. M.N. Sukhatme, Director of Herringer Bright Chemicals Pvt. Ltd, the manufacturer of the neem-based insecticide Indiara, was under pressure from Grace to sell them the technology for a storage-stable neem extract, which does not require heating or any chemical change. Sukhatme refused their offers, stating: 'I am not interested in commercialising the product.'

However, Grace eventually managed to enter into a joint venture with a firm called P J Margo Pvt. Ltd., which set up a plant in India to process neem seed for export to the US. Initially, the plant was set to process 20 tons of seed a day. They also set up a network of neem seed suppliers to ensure a constant supply of the seed at a reliable price.

In 1992, the National Research Council (NRC) in the US published a report designed to 'open up the Western world's corporate eyes to the seemingly endless variety of products the tree might offer.' According to one member of the NRC panel, 'In this day and age, when we're not very happy about synthetic pesticides, [neem] has great appeal.'

This appeal was blatantly commercial; the US pesticides market then was worth about $2 billion. Biopesticides, such as pyrethrum, together with their synthetic mimics, constituted about $450 million of this, but that figure was expected to rise to over $800 million by 1998. 'Squeezing bucks out of the neem ought to be relatively easy,' observed *Science* magazine.

Grace's aggressive interest in Indian neem production provoked a chorus of objections from Indian scientists, farmers and political activists, who said that MNCs had no right to expropriate the fruit of centuries of indigenous experimentation and several decades of Indian scientific research. This, in turn, stimulated a bitter transcontinental debate about the ethics of intellectual property and patent rights.

In April 1993, a Congressional Research Service (CRS) report to the US Congress set out some of the arguments used to justify patenting:

> *Azadirachtin* itself is a natural product found in the seed of the neem tree and it is the significant active component. There is no patent on it, perhaps because everyone recognises it as a product of nature. But ... a synthetic form of a naturally occurring compound may be patentable, because the synthetic form is not technically a product of nature, and the process by which the compound is synthesised may be patentable.

However, neither *azadirachtin*, a relatively complex chemical, nor any of the other active principles had been synthesised in laboratories; existing patents applied only to methods of extracting the natural chemical in the form of a stable emulsion or solution, methods which were simply an extension of traditional processes, used for millennia, for making neem-based products. The biologically active polar chemicals can be extracted using technology already available to villagers in developing countries, said Eugene Schulz, chair of the NRC panel. Villagers 'smash 'em [the seeds] up, soak [them] in cold water overnight, scoop the emulsion off the top and throw it on the crops'.

W R Grace's justification for patents, therefore, pivoted on the claim that their modernised extraction processes constituted a genuine innovation:

> Although traditional knowledge inspired the research and development that led to these patented compositions and processes, they were considered sufficiently novel and different from the original product of nature and the traditional method of use to be patentable.
>
> *Azadirachtin*, which was being destroyed during conventional processing of neem oil/neem cake is being additionally extracted in the form of Water Soluble Neem Extract and hence it is an add-on rather than a substitute to the current neem industry in India.

In short, the processes were supposedly novel and an advance on Indian techniques. However, the novelty existed mainly in the context of ignorance. Over the 2,000 years that neem-based biopesticides and medicines have been used in India, many complex processes were developed to make them available for specific use, though the active ingredients were not given Latinised scientific names. Common knowledge and the common use of neem was one of the primary reasons given by the Indian Central Insecticides Board for not registering neem products under the Insecticides Act, 1968. The Board argued that neem materials had been in extensive use in India for various purposes since time immemorial, without any known deleterious effects. The USEPA, on the other hand, did not accept the validity of traditional knowledge and imposed a full series of safety tests on Margosan-O.

The allegation that *azadirachtin* was being destroyed during traditional processing was inaccurate. The extracts were subject to degradation, but this was never a problem as farmers put such extracts to use as and when they needed them. The problem of stabilisation arose only when it needed to be packaged for long periods in order to be marketed commercially. Moreover, stabilisation and other advances attributable to modern laboratory technology had already been developed by Indian scientists in the 1960s and 1970s, well before the US and Japanese companies expressed an interest in them. Dr R.P. Singh of the Indian Agricultural Research Institute (IARI) said:

> Margosan-O is a simple ethanolic extract of neem seed kernel. In the late Sixties, we discovered the potency of not only ethanolic extract, but also other extracts of neem ... Work on the neem as pesticide originated from this division as early as 1962. Extraction techniques were also developed in a couple of years. The *azadirachtin*-rich dust was developed by me.

The reluctance of Indian scientists to patent their inventions, thus leaving their work vulnerable to piracy, may in part derive from a recognition that the bulk of the work had already been accomplished by generations of anonymous experimenters. This debt has yet to be acknowledged by US patentors and their apologists. The claim of the CRS report that 'the method of scattering ground neem seeds as a pesticide would not be a patentable process, because this process ... would be deemed obvious' betrayed either lamentable misjudgment or a racist dismissal of indigenous knowledge. The discovery of neem's pesticidal properties and of how to process it was by no means 'obvious'; it evolved through extended systematic knowledge development in non-western cultures. In contrast to this first non-obvious leap of knowledge, it is the subsequent minor derivatives that are 'obvious'.

W R Grace and P J Margo also claimed that their project benefited the Indian economy, by

> providing employment opportunities at the local level and higher remuneration to the farmers as the price of neem seed has gone up in the recent times because value is being added to it during its process. Over the last 20 years the price of neem seed has gone up from Rs 300 a ton to current levels of Rs 3000-4000 a ton.

In fact, the price had risen considerably higher: in 1992, Grace was facing prices of up to $300 (over Rs 8,000) per ton.

This increase in the price of neem seed turned an often-free resource into an exorbitantly priced one, with local users competing for the seed with an industry supplying consumers in the North. As the local farmer could not afford to pay what the industry did, the diversion of the seed as raw material from the community to industry ultimately established a regime in which a handful of companies holding patents would control all access to neem as raw material and all production processes. P J Margo claimed that this 'is a classic case of converting waste to wealth and is beneficial to the Indian farmer and its economy'. This statement was actually a classic example of the assumption that the local use of a product does not create wealth, and that wealth is created only when corporations commercialise the resources used by local communities.

There was a growing awareness throughout India that the commodification of neem would result in its expropriation by MNCs. When I found out that our neem had been patented, I started the neem campaign. In March 1994, around 200,000 farmers in India gathered in Delhi demanding, among other things, that the draft treaty proposed by the GATT Uruguay Round (known colloquially as 'the Dunkel Draft' after the chief negotiator, Arthur Dunkel) be translated into all Indian languages so that it could be communicated to farmers across the country. On August 15, 1994, Indian Independence Day, farmers carrying neem branches in Karnataka rallied outside the office of the District Collector in each district in the state to challenge the claims of MNCs like W R Grace demanding 'intellectual property rights'.

Their campaign was supported by many noted Indian scientists. Dr R.P. Singh expressed his 'whole[hearted] support [for the] campaign against the globalisation of the neem'. Dr B.N. Dhawan, Emeritus Scientist at the Central Drug Research Institute, maintained: 'It is really unfortunate that the benefits of all this work should go to an individual or to a company. I sincerely hope that ... the neem will continue to remain available for use by people all over the world without paying a high price to a company.' Dr V.P. Sharma, Director of the Malaria Research Centre (now NIMR), agreed: 'We have discovered the repellent action of the neem oil. There is no question of anybody else in India or outside taking a priority or patent on this aspect of neem oil. I would like this discovery to be used as widely as possible to prevent nuisance from insect pests of public health importance and in the prevention of diseases transmitted by them.' On October 2, around 500,000 farmers, along with M.D. Nanjundaswamy[4] of the Bharatiya Kisan Union (Karnataka), converged in Bangalore to voice their fears about the impending legislation, aware of the threat that GATT posed to their livelihoods by allowing multinationals to enter markets in the South at their expense.

In particular, many of them also began questioning the Dunkel Draft's call for an international harmonisation of property rights legislation. In their demonstrations, protesters carried twigs or branches of neem. Several extracts of neem had been patented by US companies, and many farmers were incensed at what they regarded as intellectual piracy. The village neem tree became a symbol of Indian indigenous knowledge, and of resistance against companies which wanted to expropriate this knowledge for their own profit.

In 1994, after the Munich-based European Patent Office (EPO) granted the patent for a neem-based biopesticide to the US Department of Agriculture (USDA) and W R Grace, I, along with Linda Bullard, the then president of the International Federation of Organic Agriculture Movements (IFOAM) and Magda Aelvoet, a Green Party leader, jointly filed a case in the EPO against the patent, challenging it on the grounds that the fungicidal properties of neem had been known in India for over 2,000 years and was not a novel product as claimed by the multinational. The IFOAM was set up in 1972 with five founding members – Lady Eve Balfour of the Soil Association of Great Britain, Kjell Arman of the Swedish Biodynamic Association, Pauline Raphaely of the Soil Association of South Africa, Jerome Goldstein of the US-based Rodale Press and Roland Chevriot of Nature et Progrès in France.

Apart from IFOAM and the Green Party, Namalvar, an organic farmer from India, and Dr Singh, a scientist from the Banaras Hindu University (BHU), joined hands in this battle against biopiracy, appearing as witnesses. Dr Fritz Dolder, Professor of Intellectual Property with the Faculty of Law at the University of Basel, Switzerland, fought the case pro bono.

Although the patent was revoked in 2000, the victory was short-lived because the USDA and W R Grace immediately filed an appeal. It took 11 years, but finally, on International Women's Day, March 8, 2005, the EPO upheld its decision to revoke the patent in its entirety.

The landmark decision upheld the values of traditional knowledge not only in India, but throughout the Global South. It was the outcome of over a decade long struggle and solidarity of committed citizens over commercial interests.

Notes

[1] A renowned biophysicist who founded the Molecular Biophysics Unit at the Indian Institute of Science, later known as the Centre of Advanced Study in Crystallography and Biophysics.

[2] A disciple of Sarala Behn, Bhatt took active part in the Chipko movement, the anti-mining and anti-liquor agitations, and the agitation against Tehri Dam from 1970 to 1990.

[3] Author of *Famine in Africa*; *Africa in Crisis*; and *Only One Earth*, among others.

[4] Scholar-activist Nanjundaswamy was India's leading advocate of farmers' rights and a vociferous anti-globalisation campaigner.

Seeds of Freedom
Towards Food Security

WHILE WORKING AT THE IIM BANGALORE IN THE EARLY 1980s, I was contacted by the eminent political scientist, Rajni Kothari, to become part of Lokayan, an initiative for bringing social movements together. Rajni was the founder of the Centre for the Study of Developing Societies (CSDS), where I met brilliant public intellectuals like Ashis Nandy,[1] Giri Deshingkar[2] and Dhirubhai (Sheth)[3] at their vibrant campus on Rajpur Road in Old Delhi. Ashis and I shared our understanding of science as a social and cultural process shaped by the worldview and objectives of the community it served. Ashis is a psychologist and I was trained in physics, but we were both aware of the violence of a reductionist science that discounted indigenous knowledge systems.

Lokayan also brought together activists like Claude Alvares and Achyut Yagnik; Claude set up the Goa Foundation with his wife, Norma Alvares, who is a lawyer and has consistently fought cases to protect Goa's coasts from the tourism industry and her forests from the mining industry. Achyut is a human rights activist who played a critical role in the Nav Nirman students' movement in Gujarat in the 1970s. In 1982, he set up SETU: Centre for Social Knowledge and Action, and I started the Research Foundation for Science, Technology and Ecology (RFSTE).

I remember meeting Rajni, Achyut and Kartikey Sarabhai[4] at Gandhi's Sabarmati Ashram in Ahmedabad and examining the details of the biggest dam project planned in India, the Narmada Dam, with a fine-tooth comb, after which we evolved a participatory study to assess its potential impact on the environment. Young activists like Medha Patkar and Bela Bhatia joined SETU and travelled the length and breadth of the Narmada Valley, leading to the birth of the Narmada Bachao Andolan. Today, Medha and Bela are major figures in environment and other social movements. Bela has been fighting for the rights of adivasis in Chhattisgarh, where corporate resource-grab has unleashed a war against tribals, and Medha continues to fight for the rights of the people displaced by the dam in Narmada Valley; through the National Alliance of People's Movements (NAPM) she is actively resisting the violation of people's rights across the country.

In 1982, Rajni invited me to join the United Nations University (UNU) Programme on Peace and Global Transformation, which he had initiated with Mushakoji, the then Vice-Rector of UNU. I was asked to develop the component that dealt with conflict over resources. This programme was aimed at evolving new knowledge paradigms based on participatory action research with diverse movements; for this, I worked with forest movements, river movements and fisherfolk movements. We did not merely document their perspectives and strategies; we created the next level of movement building.

In 1984, various farmers' movements in India blockaded the Governor's house in Chandigarh, in protest against policies regarding river water, agricultural prices and farmers' debt in Punjab. But after June 4, an essentially resource conflict narrative was communalised and made into a religious issue. This resulted in Operation Bluestar, an attack on the Golden Temple in Amritsar, and then in Indira Gandhi's assassination in October that year. I asked the UNU if I could look into the Punjab conflicts more closely and my study *The Violence of the Green Revolution* was a result of that enquiry.

Having delved into the roots of the violence of chemical agriculture in Adolf Hitler's laboratories, I G Farben (the chemical cartel that Hitler created to produce chemicals that gassed people in the concentration camps), I made a commitment to seek out nonviolent ways of farming.

In the mid-1980s, together with other activists who were concerned about the repercussions of chemical agriculture, we created a movement, Samvardhan, which we launched at Gandhi's hut in Sevagram. Dr R.H. Richharia, Banwari Lal Choudhary,[5] Claude Alvares, Korah Mathen,[6] and Kanakmal Gandhi[7] were all part of our group. It was at the Samvardhan meetings that I first met Dr Richharia, who became my seed guru.

Dr Richharia, one of India's most eminent rice scientists, used to be the director of the Central Rice Research Institute (CRRI); his research had shown that Indian peasants had, over the years, evolved 200,000 rice varieties from one wild grass, *oryza sativa*. When the World Bank wanted to take India's rice diversity to the newly-created International Rice Research Institute (IRRI) in the Philippines to appropriate our rice collections, Dr Richharia resisted. However, due to enormous pressure exerted by the World Bank, the Indian government had him removed as director of CRRI. He then went on to head the

Madhya Pradesh Rice Research Institute, but once again the World Bank had him removed by leaning on the Madhya Pradesh government. He continued his work regardless.

It was at the 'Laws of Life' conference on new biotechnologies in Geneva in 1987 that I heard what I call the Poison Cartel talk about using genetically modified organisms, or GMOs, to patent seed varieties. Activists like Pat Mooney of the ETC Group, Henk Hobbelink of GRAIN, José Lutzenberger and Anwar Fazal of the International Organisation of Consumers Union were present, as well as giants of the chemical industry like Ciba and Sandoz.

At the press conference held at the UN Headquarters in Geneva, Chakravarthi Raghavan (who had headed the Press Trust of India up until the 1975 Emergency when Indira Gandhi exiled him, and who was now covering the UN) told us that the seed and patent issue was being negotiated in the GATT negotiations, not at the UN. That was the first time I heard of GATT. After 1987, I met Raghavan whenever I visited Geneva. His book *Recolonization: GATT, the Uruguay Round & the Third World*, published by Third World Network (TWN), clearly established how corporate globalisation was a return of colonial trade. Raghavan, who I will always remember as my globalisation guru, passed away on September 28, 2021. That Geneva meeting led to my lifelong commitment to save the seed, and to monitor both GATT and globalisation. On the flight back from Geneva, I realised that the seed would be the spinning wheel of our times.

In 1987, when I commenced my project, Navdanya, I called on Dr Richharia to spend time with me in our home in Dehradun. He taught me love and respect for our indigenous seeds and for the science of breeding developed by our farmers. When my friends, the activists Ilina and Binayak Sen, wanted to start a Navdanya-like initiative in Chhattisgarh, they stayed with me for a week to learn about indigenous knowledge and seed conservation. Later, they continued Dr Richharia's work through their organisation, Rupantar. In 2018-19, when Ilina found out she had cancer, she asked me to take care of Dr Richharia's seed collection. I gave her my commitment to do so, and the Navdanya team now looks after Dr Richharia's community seed bank in Orissa. Ilina passed away in 2020, and we invited her daughter to a memorial ceremony we held to honour her life and work. The tribals in Chhattisgarh have bred a new green rice which they presented to me at our annual Bhoomi festival in October 2021.

In 1991, as part of the neoliberal reforms ushered into India, the World Bank imposed a structural adjustment programme (SAP) with stringent conditionalities that governed its loan to the government. That same year the Dunkel Draft Text of GATT was leaked – we called it the DDT. I read the intellectual property clauses in it, wrote a critique of them, and began to travel across India informing leaders in tribal communities and farmers' movements of the serious consequences if the treaty were implemented. One of the people I contacted was a Bharatiya Kisan Union leader, Mahendra Singh Tikait,[8] from Sisauli village in western Uttar Pradesh; Nanjundaswamy in Karnataka; and farm leaders in Punjab.

Agriculture is a state subject in the Indian Constitution. As the capital and chemical-intensive Green Revolution started to trap farmers in a cycle of indebtedness, they formed state unions to fight for Freedom from Debt (*Karz Mukti*) and for fair and remunerative prices that would cover their high cost of production and ensure a decent livelihood. Globalisation has allowed global corporations to hijack agriculture in India and elsewhere and undermine constitutional safeguards. We unified farmers' movements in India and globally through our movement against the unfair corporate rules of GATT and the WTO. In 1992, all the Indian farm leaders gathered in Hospet, Karnataka, in a show of strength. In 1993, 500,000 farmers came together in Cubbon Park in Bangalore. I invited every farm leader I knew across the world to participate. Later, the farmers' movement evolved as Via Campesina, the Peasant Way, and I nominated Tikait and Nanjundaswamy as the Indian representatives. In 1994, we gathered at the Red Fort in Delhi and called on the government to not sign away our farmers' future, but the WTO provisions were signed into effect in Marrakesh, Morocco, for the sake of a few jobs in Information Technology (IT), and structural adjustment has continued.

In 1992, the Earth Summit was organised in Rio de Janeiro. I worked closely with the Indian government and the TWN to protect the biodiversity of the South, and the sovereignty of our communities and countries. We ensured that Article 19.3 was introduced in the Convention on Biological Diversity to assess the impact of GMOs on biodiversity. I was invited to be a member of the expert group that created a framework for the Biosafety Protocol, signed as the Cartagena Protocol on Biosafety, now the international law on GMOs.

In 1994, I organised a meeting with the TWN of scientists and GMO activists in Penang. Eminent scientists like Brian Goodwin and Mae-Wan Ho brought the science of living organisms to activists fighting the biggest corporations with their latest toxic tools in order to strengthen the movement. I had met Mae-Wan and Brian at Teddy Goldsmith's house in London. Mae-Wan, a petite Hong Kong Chinese, was teaching genetics at the Open University and writing a book on the quantum phenomenon in living organisms; she asked me to read her manuscript because of my background in quantum physics. Her book *The Rainbow and the Worm: The Physics of Organisms* establishes how life is a self-organised system in quantum conference, from the molecule to the universe.

Brian also taught at the Open University, and he introduced me to the science of qualities. After he retired he started the Holistic Biology course at Schumacher College, where Satish Kumar had been inviting me to teach since 1991. It was while teaching at Schumacher that Brian, Martin Khor, Tewolde Egziabher and I met to plan the Penang Conference 'Beyond Reductionist Biology'.

In 1998, the chemical company Monsanto entered India illegally, without obtaining statutory approval for their GMO product. I sued them in the Supreme Court of India, and we built a national movement to keep India GMO-free. The same year, Monsanto sued Percy Schmeiser, a farmer in Canada, after contaminating his canola crop with their Round-Up Resistant GMO Canola. Percy was a gentle farmer from the Canadian Prairies whom I met while I was on a lecture tour at the University of Saskatchewan. I invited him to India to present evidence to our Parliament, which at the time was amending our patent laws, to alert parliamentarians to the consequences of patents on genes. For our part, we intervened in the Supreme Court of Canada to support him. Percy was a fellow warrior, resisting patents on seeds since 1998. He and his wife, Louise, and I travelled the world together as seed warriors. Percy passed away on October 30, 2020, at the age of 89.

Since I had busted the myth of the Green Revolution and of GMOs in the 1990s, Monsanto mounted a systematic campaign against me. They continued spreading the false narrative that organic farming does not produce enough food, and presented me with the Bullshit Award for 'making the world starve with organic' at the Rio-Plus-10 Summit in Johannesburg in 2002. I received the cowdung with gratitude and we organised a wonderful composting ceremony to celebrate the Organic. Inspired by the Monsanto award, Swedish filmmakers Per-Åke Holmquist and Suzanne Khardalian made the documentary *Bullshit*, which was released in 2005. Monsanto also attacked me for 'importing' the 'foreign' concept of organic to India. That encouraged me to start the Howard Lectures (inspired by Sir Albert Howard, who wrote *An Agricultural Testament* to honour the indigenous knowledge of our farmers) to show that India was, in fact, the source of organic farming and that organic agriculture spread worldwide from the practices of our farmers. We invited Patrick Holden, Director of the Soil Association; H.R.H. the Prince of Wales, Prince Charles; Renate Künast, the former agriculture minister of

Germany; and Masanobu Fukuoka, the Japanese farmer and philosopher who wrote *The One-Straw Revolution* as speakers.

In 1998, Monsanto also launched an attack on the world's top Lectin expert, Árpád Pusztai, whose research for the British government indicated that GMOs were causing great harm. His lab was closed and a gag order was placed on him. We organised, as scientists, to pressure UK parliamentarians to lift the gag order on him. Árpád had gone to the UK as a refugee from Hungary; following Monsanto's vicious attack, he returned to Hungary saying, 'I came to the West for freedom, and I am returning to Hungary for freedom.' We honoured both Percy and Árpád with Gandhi awards.

In 2002, we invited all the scientist warriors who had stood up against Monsanto and the Poison Cartel to honour them for their scientific integrity and their defence of nature and people.

Later, I met Dr Gilles-Éric Séralini,[9] whose pioneering work on Round-Up Resistant GMOs revealed how Round-Up was leading to organ failure and tumours in humans. His research was to be published in the journal *Food and Chemical Toxicology*, but Monsanto got wind of it, had the editor of the journal removed, and the paper was retracted. It was finally published in another journal. Subsequently, a 2015 study by the World Health Organization's (WHO) International Agency on Cancer Research identified Round-Up/glyphosate as a probable carcinogen (Group 2A), and more than 125,000 cases of victims of cancer due to Round-Up are currently pending in US courts. Bobby Kennedy Jr., son of the late Senator Robert Kennedy, has been a senior counsel in many cases, fought by the firm of Baum, Hedlund, Aristei & Goldman.

I first met Bobby when he invited me to give the keynote address at the Water-Keepers' Conference, in the Bay area, in 2013. Our common commitment to protect life on earth and the rights of people has made our paths cross, and we last met at the National Heirloom Expo in Santa Rosa, California, in 2019, where he and I addressed a panel on 'Poison-Free Food and Farming'. We last talked when he interviewed me on Bill Gates. I have been participating in the National Heirloom Expo every year since it began in 2011.

My work against the Poison Cartel and their GMOs and patents took me across the world to work with movements from Peru to Ecuador, Brazil, Argentina and Mexico; from Ethiopia to South Africa, Tanzania, Ghana and Nigeria; and to Japan, Bangladesh, Indonesia, Sri Lanka and New Zealand. Ela Gandhi, Mahatma Gandhi's granddaughter, invited me to address the South African Parliament on GMOs in 2004. Tewolde Egziabher, the environment minister of Ethiopia, invited me to his country, and we have worked closely on protecting biodiversity and shaping scientific and legal frameworks that recognise indigenous knowledge and community rights.

In Argentina, I supported the Mothers of Ituzaingó who were resisting Monsanto. I intervened in a Supreme Court case as an amicus to uphold Argentinian laws on no patents on seeds, following which the Monsanto plant was closed down. I also worked with Brazilian movements and their government on laws upholding indigenous knowledge and rights along with biosafety, as I witnessed the invasion of the Amazon by Monsanto and Cargill to grow GMO soya.

In 2016, together with 25 movements, we organised a Peoples' Assembly and a Monsanto Tribunal in The Hague to try Monsanto for crimes against nature and humanity. The following year, Monsanto disappeared, having been acquired by Bayer.

Our Peoples' Campaign against the WTO in India continues. S.P. Shukla, India's erstwhile ambassador to GATT, was its convenor, our former prime minister, V.P. Singh, was the chair, and the campaign has had a huge influence on negotiations, on Parliament, and in the writing of laws. This is why India has a patent law that does not allow patents on seeds, animals and plants, and a Farmers' Rights Act which recognises the right of farmers to save, exchange, improve, breed and sell seeds.

*****

The Monsanto Tribunal and Peoples' Assemblies that took place are important examples of mobilising against a powerful multinational – or rather, transnational – corporation, demanding accountability from it in a people's court. I think it's important to recount this experience. Corporate crimes have become visible everywhere, with corporations becoming bigger, claiming absolute power, absolute rights and absolute immunity, while deploying more violent tools against nature and people. In October 2016, we organised a Monsanto Tribunal and Peoples' Assembly to put the Poison Cartel on trial at The Hague. Whereas the Tribunal both synthesised the existing crimes and violations for which Bayer-Monsanto is in courts across the world – in India, Europe, the U.S., Mexico and Argentina – and expanded the scope of criminal activity to include the crime of ecocide, a violation of the rights of nature, the Peoples' Assembly did not just take stock of past and present crimes, it looked at future crimes with the aim of preventing them. The Peoples' Assembly in The Hague, a gathering of movements, seed savers, seed defenders, farmers and civilians followed on the heels of several such assemblies organised by local communities the world over.

When Monsanto became Bayer-Monsanto, Syngenta merged with Chem China, and Dow merged with Dupont, movements from India, China, Germany and Switzerland began challenging these mergers by organising Peoples' Assemblies and planning future action against them. The chemical corporations had hoped to take over all seed production by 2000 through GMOs, patents, mergers and acquisitions, but most seeds are not genetically modified, and most countries do not recognise seeds and plants as corporate 'inventions', hence patentable.

The Monsanto Tribunal was the culmination of thirty years of scientific, legal, social and political mobilising by movements, concerned citizens and scientists. The organisations that supported the Tribunal and Assembly in The Hague included Arrêtons l'Ecocide – End Ecocide on Earth; ASEED; ATTAC; Biovision Foundation for Ecological Development; CRIIGEN; Forum Civique Européen; Générations Futures; Gezonde Gronden; GRAIN; IFOAM-Organics International; La Via Campesina; Millennium

Institute; Navdanya International; Netwerk Vitale Landbouw en Voeding (Network Vital Agriculture and Nutrition); Netzfrauen; Organic Consumers Association; Pesticide Action Network Europe; Regeneration International; SAG/ StopOGM-Alliance suisse pour une agriculture sans génie génétique; Save Our Seeds; and Supermacht.

We chose The Hague, where the International Criminal Court (ICC) is located, for the Tribunal and Assembly because the ICC was set up in 2002 to investigate war crimes and genocide linked to conflicts, and to shape the future of jurisprudence. With the international court widening its remit to look at the destruction of the environment as a violation of people's rights to their resources, and prioritising crimes that result in the 'destruction of the environment', 'exploitation of natural resources', and the 'illegal dispossession' of land – its remit also included an explicit reference to land-grabbing – movements in defence of human rights and the rights of the earth gathered to demonstrate that ecological destruction is a war against the earth, that ecocide is a crime against nature, and crimes against nature are connected to crimes against humanity.

While courts can investigate the crimes of the Poison Cartel, people have the power to bring change through self-organisation, through Peoples' Assemblies worldwide. From the hundreds of assemblies held everywhere we launched a global campaign to liberate our seeds and soils, our communities and societies, our planet and ourselves, from the rule of the Poison Cartel.

The Monsanto Tribunal was convened on October 15, and comprised five judges – Eleonora Lamm, who had served as human rights director for the Supreme Court of Justice of Mendoza in Argentina; Françoise Tulkens of Belgium, who had been a judge in the European Court of Human Rights and was vice-chair of the Scientific Committee of the European Union Agency for Fundamental Rights (FRA); Steven Shrybman of Canada, who was a member of the Boards of the Council of Canadians and the Institute for Agriculture and Trade Policy; Jorge Fernández Souza of Mexico, who was a judge at the Court of Administrative Litigation of Mexico City; and Dior Fall Sow of Senegal, a consultant to the ICC and former Advocate General at the International Criminal Tribunal for Rwanda.

The enquiry was based on six key issues: the right to a healthy environment; the right to food; the right to the highest attainable standard of health; the freedom indispensable for scientific research; whether Monsanto was complicit in the commission of a war crime by providing materials to the US army in the context of Operation Ranch Hand launched in Vietnam in 1962; and whether Monsanto's activities constitute a crime of ecocide.

Over two days, the judges heard the testimonies of 28 witnesses and public health and environment experts from across the world, who testified how Monsanto had violated human rights and committed crimes against the people, and the planet, by aggressively promoting its products, lobbying with politicians and attacking independent scientists.

In the end, the Tribunal concluded that Monsanto had indeed violated the human rights to food, health, a healthy environment and the freedom indispensable for independent scientific research. Stating that human rights and environmental laws are undermined by corporate-friendly trade and investment regulation, it strongly called for the need to recognise 'ecocide' as a crime under international law.

Alongside the proceedings of the Tribunal, the Peoples' Assembly brought together several scientists, activists and leaders of key organisations like Navdanya International, Biovision, IFOAM, GM Watch, Via Campesina and Pesticide Action Network, among others, for experience-sharing, discussions and workshops. Among the speakers at the Assembly were André Leu, president of IFOAM; Ronnie Cummins of the Organic Consumers Association; Hans Herren, chair of IAASTD and Biovision; activist and poet Nnimmo Bassey of the Health of Mother Earth Foundation; Claire Robinson, editor at *GMWatch.org*; scientist Shiv Chopra; Percy Schmeiser; and myself.

*****

When I was in Florence in 2002 for an Environment Day lecture for the Ministry of the Environment in Tuscany, Italy, the president, Claudio Martini, asked to meet me. I was to leave by an early morning flight the next day. He came to my hotel before I left and invited me to work with him and his government on the San Rassore Dialogues on Alternatives to Economic Globalization. Caroline Lockhart, who had worked for the UN, was invited from New York to coordinate the process. Prince Hassan of Jordan, Al Gore and Teddy Goldsmith came as speakers.

The International Commission on the Future of Food and Agriculture (ICFFA), chaired by Martini and myself,

grew out of the San Rassore Dialogues. I was given a free
hand by the president to organise the Commission. Bringing
together diverse strands of movements working on the same
issue has always been my strategy for building movements,
and it has been my philosophy as well.

For our first meeting we invited Jerry Mander of
the International Forum on Globalisation (IFG); Teddy
Goldsmith; Carlo Petrini, who founded the Slow Food
Movement; and IFOAM, the organic movement, to
participate. It was Jerry's idea to call our declarations
Manifestoes. At the end of the meeting, Carlo asked how we
could bring them to life. I suggested we do so by bringing
together the peasants and food producers of the world; and
so Terra Madre, the biggest food event, was born. It was
organised by Slow Food and ran parallel to the Slow Food
Fair, Salone del Gusto, in Turin, in 2006. The *Manifesto on
Seed* was released at Terra Madre, and I was vice president
of Slow Food for three years. Navdanya participated in the
Presidium Programme of Slow Food through the Basmati
Presidium and the Mustard Presidium.

The *Climate Manifesto* was released at the Copenhagen
Climate Change Conference in 2009, in the presence of
more than one thousand movements. After the 2008 financial
crisis we wrote the *Knowledge Manifesto*. In the Year of the
Soil, 2015, as refugees sank in boats in the Mediterranean,
we connected the soil, climate and refugee crises in the
manifesto *Terra Viva: Our Soils, Our Commons, Our Future*.

The Commission was slated to meet and prepare a
Manifesto on the *Economy of Care* at the end of March 2020,
when the COVID lockdowns were announced, but we have

continued our work in international movement-building on the rights of the earth and the rights of people in this new context. Building on my book *Oneness vs. the 1%,* we invited all movements working on different aspects of the Gates Empire to write a collective report, *Gates to a Global Empire*, which has created a new awareness that the world's billionaires are running the show. Gates and the World Economic Forum (WEF) have taken over the Food Summit from the FAO; now they are trying to hijack COP26.

Ever since the ICFFA was created in 2003 in Tuscany, it has published manifestoes on issues of critical importance to the future of the planet. These include *Future of Food* (2003); *Future of the Seed* (2006); *Climate Change and the Future of Food Security* (2008); *Future of Knowledge Systems* (2009); *Terra Viva* (2015) and *Food for Health* (2019).

The manifestoes present a synthesis of the work and ideas espoused by hundreds of international organisations and thousands of individuals, actively seeking to reverse the present dire trend towards the industrialisation and globalisation of our food, our health and our knowledge systems. While they critique the dangerous directions of the moment, more importantly, they set out a practical vision and share ideas and programmes to enable local communities and governments to create systems that are more socially and ecologically sustainable by putting food quality, food safety and public health above corporate profit.

After the collapse of the global financial system brought on by the Wall Street disaster, it was southern Europe that bore the brunt, with the economies of Greece, Italy, Spain and Portugal being restructured to pay their debts. Agencies and organisations associated with me in Tuscany were shut down, but I promised the people who worked with me that I would continue with our common endeavour if they formed an association that could do so. This is how Navdanya International came into being, and I was nominated its president. Caroline Lockhart, who was ICFFA's coordinator, became vice president. Board members include Piero Bevilacqua[10] and Salvatore Ceccarelli[11] of Italy, Nnimmo Bassey of Nigeria, Bernward Geier[12] of Germany and Debbie Barker[13] of the US.

Navdanya International was founded in Italy in 2011 to strengthen Navdanya's global outreach in its mission to protect nature, the earth's biodiversity and people's rights to seed and food, as well as to protect farmers' rights to save, exchange and evolve seeds, and to protect indigenous knowledge and culture.

Navdanya was established when I realised that globalisation was changing India's agriculture and indigenous seeds were disappearing fast, but the work that was required could not be done by me alone. I wrote to a few funders, including Peter Rottach at Bread for the World, who immediately understood the urgency of the task. Kamla Bhasin, who worked then in the Freedom from Hunger Campaign of the FAO, helped implement Navdanya's initial seed-saving work through her office. In 1994, we set up a seed bank in the Doon Valley which has evolved into a leading learning and research centre on seed saving, biodiversity conservation, agroecology and regenerative agriculture. Over time, we have created hundreds of community seed banks, facilitated thousands of seed savers, and spread the seed freedom movement worldwide. Our grassroots movement, Shakti, is creating hope in the midst of climate disasters, health in times of COVID, nonviolent power in creative form in the face of accelerated violence threatening the collapse of ecosystems and societies.

Through my work on seeds and food, I have met some of the most courageous and committed people in the world who are passionate about protecting the earth, her biodiversity, and the right to food and health for all.

Wendell Berry, the US farmer who wrote the book *The Unsettling of America*, informed me that America's farmers were unsettled by corporate industrial agriculture just as Punjab was unsettled by the Green Revolution. Wendell gave me an award in the memory of his father during the 25th anniversary of his book in 2012, and wrote the introduction to *The Vandana Shiva Reader* published in 2015.

I worked with Winona LaDuke, a Native American activist who ran as vice presidential candidate with Ralph Nader in 1996 and 2000, to help her fight the patent on native wild rice, based on our fight against the biopiracy of basmati by RiceTec in 1998. Miguel Altieri[14] and Ignacio Chapela[15] of Berkeley became good friends in promoting ecological agriculture and speaking the scientific truth about the ecological impact of GMOs. Hans Herren, chair of the International Assessment of Agricultural Knowledge, Science and Technology for Development (often referred to as the IPCC of Agriculture), was a close associate. André Leu, who had served as president of IFOAM for eight years, Ronnie

Cummins with whom I started the Global Days of Action on GMOs, and 60 others from 21 countries met in Costa Rica in June 2015 to draw up a blueprint on the common goal of reversing global warming and ending world hunger. Our independent work showed that both objectives could be met by obeying the ecological laws of nature. I wrote *Soil Not Oil* in the lead-up to the Copenhagen Climate Change Conference in 2009, and *Who Really Feeds the World?* for the Milan Food Expo in 2015. Others, too, had arrived at similar conclusions. We agreed to put our energies together and build the regeneration movement, which is how Regeneration International was born in 2015.

Food directed me to health, and to an amazing community of health practitioners, doctors, and ayurveda experts. At the national level, Navdanya and the Trivandrum-based Centre for Innovation in Science and Social Action (CISSA) created Annam, the movement of food for health. Recently, Dr Gangadharan of CISSA and I wrote a book on *The Two Futures of Food, Health and Humanity*, and globally, we wrote a Manifesto on *Food for Health*.

Reclaiming the seed

Seed is the source of life. Seed creates, recreates, regenerates, and multiplies itself.

Seed is *autopoiesis,* the poetry of life, written by the orchestra of life, in interconnected autonomy; it is a self-organised interbeing, self-regulating and coherent, from the molecule to the cell, to the organism as a whole. It holds the past of cultural biology and evolution within itself. It enfolds the future potential of unfolding evolution in diversity, vibrancy, resilience.

Seed is neither a machine nor an invention. The abundance and diversity of seeds have been co-created and co-evolved by small farmers, especially women, over centuries. The chemical industry takes farmers' seeds for free and breeds them into seeds that respond to chemical inputs, or genetically engineers them with toxic genes and patents them, making them non-renewable. Patents on seeds and criminalising seed-saving and seed-exchange violate farmers' rights to sow, save, exchange, breed, and sell their seeds.

At the 'Laws of Life' conference in Geneva that I mentioned earlier, the chemical industry predicted that by the year 2000, there would be five companies controlling seed, that all seed would be genetically engineered and patented, and that it would be illegal for farmers everywhere to save and share seeds. GATT, which became WTO, would be the legal instrument to impose what I call Seed Imperialism.

Reclaiming the commons of our seeds has been my life's work for more than three decades now. Inspired by Gandhi, we started the Navdanya movement with a Seed Satyagraha, and declared,

> *Our seeds, our biodiversity, our indigenous knowledge are our common heritage. We receive our seeds from nature and our ancestors. We have a duty to save and share them, and hand them over to future generations in their richness, integrity, and diversity. Therefore we have a duty to disobey any law that makes it illegal for us to save and share our seeds.*

> *Through imposing laws related to Intellectual Property Rights and patents on seed which deny farmers and gardeners the right to save and share seed, our diversity, freedoms and right to health are being criminalised.*
>
> *Higher laws that flow from the laws of the earth, reaffirmed by the laws of our humanity, compel us to question and resist the imposition of laws based on uniformity as an instrument of control being forced upon our diversity as peoples, cultures and other species, which we have a duty to protect and defend.*
>
> *We do not recognise any laws created by corporate interests that interfere in our duty to save and share good seed so that the generations to come are as fortunate as we have been in receiving these gifts of diversity and nourishment.*
>
> *We will not obey or recognise any law that criminalises our time-tested seeds.*
>
> *This is our Seed Satyagraha.*

Our Seed Satyagraha was launched as a national movement in 1991. When attempts were made to make it illegal for farmers to save seed in Europe, in the US, and in many other countries, the Seed Satyagraha became global.

In May 2006, Navdanya undertook a Seed Pilgrimage (Bija Yatra) to stop farmers' suicides and create an agriculture of hope, to reclaim our seed and food sovereignty. The Yatra started from Gandhi's ashram in Sevagram, Maharashtra, and ended on May 26 in Bangalore. It covered Amravati, Yavatmal and Nagpur in Vidarbha region of Maharashtra; Adilabad, Warangal, Karimnagar and Hyderabad in Andhra Pradesh; and Bidar, Gulbarga, Raichur, Hospet, Chitradurga and Bangalore in Karnataka. These are the regions where farmers have been locked into dependence on corporate seed supply in order to grow cash crops integrated into world markets, leading to a collapse in farm prices due to subsidies amounting to $400 billion in rich countries.

In India, the seed sovereignty movement was successful in defending the freedom and integrity of the seed, and of farmers' rights and farmers' freedom in law. Article 3j of the Indian Patents Act clearly states that plants, animals and seeds are not inventions, hence not patentable: plants and animals in whole or in any part thereof other than micro organisms; but including seeds, varieties, and species are essentially biological processes for production or propagation of plants and animals. Article 39 of the Indian Plant Variety Protection and Farmers' Rights Act states, 'A farmer shall be deemed to be entitled to save, use, sow, re-sow, exchange, share or sell his farm produce, including seed of a variety protected under this Act, in the same manner as he was entitled to before the coming into force of this Act.' It was this Article that prevented Pepsi from suing four Indian potato farmers for Rs 10 million each.

In Europe, 180,000 citizens have submitted a petition, 'No Patents on Seed', which implies no patents on seeds via genome sequence information. Local laws in Brazil, Argentina and Mexico also prohibit patents on seed.

Through patents and their attempts to make it illegal for farmers to use their seeds, the chemical giants began their moves to control seeds worldwide; over the last two decades four giant chemical corporations – Bayer-Monsanto, Dow-Dupont, Syngenta-Chem China and BASF – have taken over control of 60 per cent of the world's seed supply. Bill Gates is now leading the Gold Rush to control seeds through digital colonisation and Big Data, and is building his empire in partnership with Bayer-Monsanto, Syngenta and the Poison Cartel.

The Iraq War was also a seed war. On April 26, 2004, Paul Bremer, the administrator of the Coalition Provisional Authority (CPA) in Iraq passed Iraqi Order 81, prohibiting Iraqi farmers from using native seeds. Article 14 of the Order states: 'Farmers shall be prohibited from re-using seeds of protected varieties.' This was a declaration of war against farmers of the fertile crescent, and it is why I joined hands with Dr Brian John[16] and Wafaa' Al-Natheema,[17] and gave a call to movements to celebrate April 26 as International Seed Day in Defence of Seed Freedom:

> Therefore, organisations, activists, organic food advocates, farm owners and farmers around the world are joining hands to advocate for patent-free seeds and biodiversity and to educate about the criminal practices by agricultural corporations and how their unjust laws have and will affect the future of agriculture.

India's Green Revolution was one example of the deliberate destruction of diversity, and new biotechnologies repeat and deepen these tendencies, rather than reverse them. Further, such technologies, combined with patent monopolies, threaten to transform the diversity of life forms into mere raw material for industrial production and limitless profit. They are simultaneously threatening the regenerative freedom of diverse species, as well as the free and sustainable economy of small peasants and producers, based on nature's diversity and its utilisation. The seed, for example, reproduces itself and multiplies. Farmers use seed both as grain and as the next year's crop. Seed is free, both in the ecological sense of reproducing itself, as well as in the economic sense of reproducing farmers' livelihoods. The seed's freedom, however, is a major obstacle for seed corporations. If a market for seeds has to be created, the seed needs to be transformed materially, so that its reproducibility is blocked and its status can be changed legally. Instead of being the common property of farming communities, it becomes the patented private property of seed corporations.

As my involvement in these issues grew, the seed started to take shape as the site and symbol of freedom in an age of manipulation and monopoly. Ethically and ecologically, unrestrained biotechnology development creates new tools for manipulation, and patents offer new means for monopoly ownership of that which is free by its very nature. I was reminded of Gandhi's spinning wheel, which became such an important symbol of freedom – not because it was big and powerful but because it was small and could come alive as a force of resistance and creativity in the smallest of huts and the poorest of families. Its power lay in smallness. The seed, too, is small. It embodies diversity. It embodies the freedom to stay alive. And it is still the common property of small farmers in India. Seed freedom goes far beyond freedom from corporations for the farmer – it signifies the freedom of diverse cultures from centralised control. In the seed, ecological issues combine with social justice. The seed can play the role of Gandhi's spinning wheel in this period of recolonisation through 'free trade'. The native seed has become a symbol of resistance against monocultures and monopoly rights. The shift from uniformity to diversity respects the rights of all species, and is sustainable. Diversity is also a political imperative because it demands decentred control. Diversity as a way of thought and a way of life is what is required to transcend the impoverished monocultures of the mind. The conservation of diversity is the commitment to let alternatives flourish in society and in nature, in economic systems and in knowledge systems. Cultivating and conserving diversity is no luxury; it is a survival imperative, and a precondition for the freedom of all, the big and the small. In diversity, the smallest element is significant; allowing the small to flourish is the real test of freedom – in the life of an individual, the life of an organisation, the life of a society and the life of this planet. It is this interdependence between diversity, decentredness and democracy which needs to be conserved and cultivated.

Seed and food the world over have been shaped by millions of years of nature's contribution and centuries of women's intelligence, skill, hard work and perseverance. Today, women are again in the vanguard of defending seed freedom and food sovereignty in the context of globalisation which, worldwide, has facilitated corporate grab through patents on seeds associated with genetic engineering; they have done this via a mechanistic paradigm of biology and agriculture, and through a reductionist paradigm of the economy. Women as activists, scientists and scholars are at the forefront of shaping new scientific and economic paradigms to reclaim seed sovereignty and food security across the world. They are leading movements to change both practice and paradigm: how we grow and transform our food. As seed keepers and food producers, as mothers and consumers, they are engaged in renewing a food system that is better aligned with the ecological processes of the earth's renewal, the laws of human rights and social justice, and the means through which our bodies stay well and healthy.

The industrial paradigm of agriculture

The industrial paradigm for food production is clearly no longer viable. It is unviable because it emerged from labs producing tools for warfare, not from farms and fields producing food and nourishment. It has its roots in war; an industry that grew by making explosives and chemicals for war remodelled itself as the agrichemical industry when the major 20th century wars ended. Explosives factories started making synthetic fertilisers, war chemicals began to be used as pesticides and herbicides. Whether it is chemical fertilisers or chemical pesticides, they are designed to kill. That is why thousands were killed in India, in Bhopal on December 2, 1984, and hundreds of thousands continue to be maimed because of leaks from a pesticide plant owned by Union Carbide (now Dow Chemicals). That is also why chemicals like Round-Up/glyphosate are being implicated in new disease epidemics by scientists like Prof. Seneff of MIT, who identify the processes through which these chemicals cause harm.

In 1984, because of the Bhopal disaster and extreme violence in Punjab – projected as the granary of India – I decided to study why agriculture had become so volatile. The Green Revolution, introduced initially in Punjab, led to the inclusion of chemicals in Indian agriculture. And although its main advocate, Norman Borlaug, was awarded the Nobel Peace Prize in 1970, by 1984 Punjab had become a land of war, not peace, and peace had broken down because the sustainability of water, the health of soils and of people were all being undermined by the chemicals which drove the 'Green Revolution'. Poisons contributed to a cancer epidemic in the state, with a 'cancer train' taking cancer victims from Bhatinda to Rajasthan for treatment.

The violence of Punjab and Bhopal propelled me into dedicating my intellectual and activist energies to creating a nonviolent paradigm for food and farming. Collectively, across the world, we have forged a new paradigm for agriculture referred to as agroecology. Agroecological systems produce more and better food, and return higher incomes to farmers.

The chemical push reoriented agriculture towards toxicity and corporate control. Instead of working with ecological processes and taking the well-being and health of the entire agroecosystem, with its diverse species, into account, agriculture was reduced to an external input system adapted to chemicals. Instead of recognising that farmers have been breeders over millennia, giving us the rich agro-biodiversity that is the basis of food security, breeding was reduced to breeding uniform industrial varieties that respond well to chemical inputs. Instead of small farms producing diversity, agriculture became focused on large monoculture farms producing a handful of commodities. Correspondingly, human diets shifted from having 8,500 plant species to about eight globally traded commodities. The scientific paradigm was also transformed. Instead of encouraging a holistic approach, the practice of agriculture was compartmentalised into fragmented disciplines based on reductionism.

Just as gross domestic product (GDP) fails to measure the real economy and the health of nature and society, the category of 'yield' fails to measure the real costs, the real outputs of farming systems. As the UN observed, the so-called High Yielding Varieties (HYVs) of the Green Revolution should, in fact, be called High Response Varieties, as they have been bred for chemicals; they are not High Yielding in and of themselves. The narrow measure of 'yield' propelled agriculture into deepening monocultures, displacing diversity and eroding natural and social capital. The social and ecological impact of this broken-down model has pushed the planet and society into deep crisis.

- Industrial monoculture agriculture has pushed more than 75 per cent of our agro-biodiversity and 93 per cent of crop biodiversity to extinction.
- 75 per cent of bees have been killed because of toxic pesticides. Einstein had cautioned, When the last bee disappears, humans will disappear.
- 75 per cent of the water on the planet is being depleted and polluted for chemical-intensive industrial agriculture. The nitrates in water from industrial farms are creating 'dead zones' in our oceans.
- 75 per cent of land and soil degradation is caused by industrial farming.
- 50 per cent of all greenhouse gas emissions responsible for climate change come from a fossil fuel, chemical-intensive globalised industrial system of agriculture. Fossil fuels, used to make fertilisers, run farm machinery,

and transport food thousands of miles from where it is grown contribute to carbon dioxide emissions. Chemical nitrogen fertilisers emit nitrogen oxide which is 300 per cent more destabilising for the climate than carbon dioxide, and factory farming is a major source of methane.

While this ecological destruction of natural capital is justified in terms of 'feeding people', the problem of hunger has only grown. When the focus of agriculture is the production of commodities for trade instead of food for nourishment, hunger and malnutrition are the outcome. Only 10 per cent of corn and soya grown in the world is used as food; the rest goes for animal feed and biofuel. Commodities do not feed people, food does.

A high cost external input system is artificially kept afloat with US $400 billion as subsidies – that is more than US $1 billion a day. The so-called 'cheap' commodities have a very high cost financially, ecologically and socially. Industrial, chemical agriculture displaces productive rural families; it is also debt-creating, and debt and mortgages are the main reason for the disappearance of the family farm. In extreme cases, as in the cotton belt of India, debt created by the purchase of high-cost seed and chemical input has pushed more than 127,000 farmers to suicide in a little over a decade.

The false argument that GMOs are needed in order to increase food production to feed growing populations is a desperate attempt to extend the life of a failing paradigm. A series of myths are formulated to bolster this argument.

Living organisms, including seed, are self-organised, complex systems. According to Mae-Wan Ho, they adapt and evolve, and are 'fluid' at the level of the genome. Genes are influenced by the environment, as the new discipline of epigenetics shows. Genetically engineered seed and food are being promoted as a technological miracle for feeding the world and ending malnutrition and hunger. However, after twenty years of commercialisation, all the promises of the GMO miracle have been discredited. The following GMO myths need to be challenged and disproved.

Myth 1: GMOs are an 'invention' of corporations, and therefore can be patented and owned. Living organisms, including seeds, thus become the 'intellectual property' of the GMO industry. Using these property rights, corporations can forcibly prevent farmers from saving and sharing seeds, and can collect royalties on their patented products. A Monsanto representative is on record stating that his company wrote the intellectual property agreement of the WTO. He added that they were the 'patient, diagnostician, physician'; they defined the problem – farmers save seeds – and offered a solution: saving seed should be made illegal.

The claim to invention is a myth because genetic engineering does not create a plant or an organism – it is merely a tool to transfer genes across species. Living organisms are self-organising, self-replicating systems. Unlike machines, they cannot be engineered. There are only two ways of introducing genes from unrelated species: one is the use of a gene gun, the other is through plant cancer. Just as a mover of furniture is not the maker or owner of the house to which the furniture is moved, the GMO industry is merely the mover of genes from one organism to another, not the creator or inventor of the organism, including seeds and plants.

Through the false claim of 'invention' and 'creation', the GMO industry is appropriating millions of years of nature's evolution and thousands of years of farmers' breeding.

Myth 2: Genetic engineering is more accurate and precise than conventional breeding. All breeding has been based on breeding within the same species: rice is bred with rice, wheat with wheat, corn with corn.

The tools of genetic engineering allow the introduction of genes from unrelated species into a plant, and include genes from bacteria, scorpions, fish, and cows. The introduction of genes from unrelated species is a blind technology, neither accurate nor precise. When genes are introduced into the cells of a plant using a gene gun, it is not known if the cell has absorbed the gene or not. That is why every GMO also uses an antibiotic-resistance-marker gene to separate cells that have absorbed the gene from those that have not. This means that every GMO in food has antibiotic resistance genes which can mix with bacteria in the human gut and aggravate the crisis of antibiotic resistance we are currently facing.

Further, since the introduced gene does not belong to the organism, genes from virulent viruses are added as 'promoters' to express the trait for which genes have been introduced. These additional transformations are evidence of the unreliability and inaccuracy of the gene transfer technology. Moreover, nothing is known about what these genes do when they enter our body as food. In the case of herbicide-tolerant crops like Round-Up Ready soya and corn, the combination that needs to be considered for its impact on the environment and health is both Round-Up/glyphosate and the new genes in the food crop.

Myth 3: GMOs are just like naturally occurring organisms, and are therefore safe. This is inconsistent with myth 1. To establish ownership, the GMO industry claims novelty. To avoid responsibility for adverse impact, it claims naturalness. I have called this 'ontological schizophrenia'. GMOs have an impact on the environment, on our health, and on farmers' socio-economic status; that is why we have an international UN Biosafety Protocol.

The unscientific principle of 'substantial equivalence' has been institutionalised in order to avoid research on biosafety. Substantial equivalence assumes that the GMO is substantially equivalent to the parent organism. This leads to a 'don't look, don't see, don't find' policy, and not having looked for impacts, GMOs are declared 'safe'. Ignorance of impacts, however, is not proof of safety. There is as yet no proof of safety.

Myth 4: GMOs are based on cutting-edge science; GMO critics are 'antiscience'. Genetic engineering is based on an obsolete paradigm of genetic determinism, a linear and deterministic flow of information from genes, which are called 'master molecules', to proteins. Francis Crick called this the 'central dogma' of molecular biology. Genetic determinism assumes that genes are atoms of biological determinism, with one gene carrying one trait, and determining the traits in an organism. But these are assumptions that come from the idea of control and domination; this is patriarchal ideology, not science.

Cutting-edge science teaches us that these assumptions are false. Genes are fluid, not fixed. Each gene contributes to multiple traits; each trait is an expression of many genes acting in concert. As Richard Lewontin states in *The Doctrine of DNA*:

> DNA is a dead molecule, among the most non-reactive chemically inert molecules in the world. It has no power to reproduce itself.... When we refer to genes as self-replicating, we endow them with a mysterious, autonomous power that seems to place them above the more ordinary materials of the body. Yet if anything in the world can be said to be self-replicating, it is not the gene, but the entire organism as a complex system.

On the basis of the latest and independent science, leading scientists across the world have contributed to the new science of biosafety, while corporate ideology, parading as science, has launched a brutal, violent and unscientific attack on every scientist speaking the truth about GMOs on the basis of detailed scientific research on the impact of GMOs; this includes Dr Árpád Pusztai, Dr Ignacio Chapela, Dr Eric Seralini and myself.

Myth 5: GMOs increase yields and are the answer to world hunger. Genetic engineering as a tool for the transfer of genes is not a breeding technology; it does not contribute to breeding high-yield crops. Yields come from conventional breeding; all that genetic engineering does is add a Bt toxin gene or a gene for herbicide tolerance, an antibiotic marker and virus genes. These do not increase food production, but they do contribute to the production of risks from toxins and antibiotic resistance.

Even the argument that GMOs increase yield indirectly by controlling weeds and pests is incorrect because rather than controlling weeds and pests, Bt GMOs *have contributed to the emergence of new pests and super-pests resistant to the Bt toxin*, and herbicide-resistant crops have led to super-weeds that are resistant to Round-Up. Hence, new GMOs have now been developed which are resistant to 2,4D, an ingredient of Agent Orange.

Myth 6: GMOs reduce chemical use and are therefore environmentally beneficial. Two applications of genetic engineering account for most commercial planting, Bt crops and Ht crops. Herbicide-tolerant crops account for 63 per cent of the cultivation of GM crops. Bt crops have led to an increase in pesticide use because of new pests and pest resistance in the boll worm. As the Directorate of Plant Protection shows, pesticide use has increased with the increase of Bt cotton cultivation. Anyway Ht crops are designed to make crops resistant to herbicide spraying.

Myth 7: GMOs promote free choice. The myth of 'free choice' begins with 'free market' and 'free trade'. When five transnational corporations control the seed market, it is not a free market, it is a cartel. When corporations write the rules of 'free trade', it is corporate dictatorship, not free trade. When enforcing patents and IPR laws written by themselves, corporations prevent farmers from saving seed; it is not 'free choice', it is seed slavery.

In India, Monsanto has locked local seed companies into licensing agreements to *only sell Bt cotton*. The labels have different names, but they are all 'Bollgard', Monsanto's Bt cotton. This is illusionary 'free choice'; the reality is seed monopoly.

When corporations spend millions to prevent the labelling of GMOs and deny citizens the Right to Know and the Right to Choose, free choice is stifled.

One agriculture, one science: capitalist patriarchy's domination and control

On July 22, 2014, the International Crops Research Institute for the Semi-Arid Tropics (ICRISAT) announced the formation of 'One Agriculture, One Science', an international partnership across India, Africa and the US launched at an expert consultation workshop jointly organised by ICRISAT, the University of Florida, Michigan State University, and Iowa State University in Gainesville, Florida. It was attended by experts from land-grant universities in the US; the Indian Council of Agricultural Research (ICAR); the Alliance for a Green Revolution in Africa (AGRA); the Regional Universities Forum for Capacity Building in Agriculture (RUFORUM – a consortium of 42 universities in 19 African countries); the US Department of Agriculture; the US Agency for International Development; and the Consultative Group on International Agricultural Research (CGIAR), a consortium of 15 international agriculture research centres.

While 'One Agriculture, One Science' is falsely promoted as a new and innovative project, it is actually a repeat of how the Green Revolution was launched in the 1960s. Then, too, the US land-grant universities were involved in training our scientists in the chemical paradigm, displacing the diversity of agriculture systems adapted to different ecosystems. Then, too, the US Department of Agriculture and USAID were involved in pushing the Green Revolution. Then, too, the World Bank–governed CGIAR centres played a key role in promoting it – were, in fact, created to launch it. The International Maize and Wheat Improvement Centre (CIMMYT) in Mexico was created to introduce the Green Revolution in maize and wheat; the International Rice Research Institute (IRRI) in the Philippines was set up to spread the Green Revolution in rice. The ICRISAT was established later and is the CGIAR centre specialising in semi-arid crops.

The only difference between the 1960s and today is that Big Money and Big Ag are directly pushing a monoculture to create monopolies for profits through ownership of seeds and the sale of chemicals. The AGRA was launched by Bill Gates and his foundation – the Gates Foundation owns 500,000 shares in Monsanto, indicating that philanthropy and business merge when Big Money meets Big Ag. Bayer-Monsanto is not just the biggest seed corporation today, it has bought up the biggest climate data corporation, Climate Corp, and a soil data corporation, Solum, which it has renamed Granular.

The objective is to overwhelm farmers with Big Data and make them dependent for every aspect of farming – seed, soil, climate data – all of which become commodities that the company sells and farmers have to pay for. But Big Data is not knowledge which comes from experience, interconnectedness, and participation. Big Data from One Corporation contributes, as we have written in the ICFFA's *Future of*

Knowledge Systems Manifesto, to 'information obesity'. It is also a means of control. Locking in 42 African universities that work across diverse cultures and in diverse ecosystems – desert and rainforest, mountain and coast – into 'One Agriculture, One Science', for example, is a recipe for impoverishing and enslaving Africa, intellectually and economically.

There are three major reasons why this grand project will aggravate, instead of solve problems in agriculture. The branding exercise that resulted in 'One Agriculture, One Science' is a failure in itself. For 'experts' to believe that different climates, different ecosystems and different cultures can be prescribed 'One' solution is laughable. Either they are not aware that the rainfall in Cherrapunji (India) is different from the rainfall in Oaxaca (Mexico), that it's hotter in Maharashtra than in Oregon, or they simply don't care whether a farmer's crop fails or succeeds as long as they have extracted every last dollar, rupee or rand from him/her.

The 'One Agriculture' push by Big Ag corporations ignores the findings of all UN agencies, including of the International Assessment of Agricultural Knowledge, Science and Technology for Development (IAASTD), a team of 400 scientists that has been working for over six years. The IAASTD report states: 'We must look to small-holder, traditional farming to deliver food security in Third World countries, through agroecological systems which are sustainable. Governments must invest in these systems. This is the clear evidence.'

'One Agriculture, One Science' ignores evidence of the failure of the Green Revolution chemical monocultures and the success of diverse agroecological systems in addressing hunger while protecting the planet. It is a call against diversity and will further erode the ecological foundations of agriculture, leaving the world's food systems at the mercy of billionaires and big corporations. It is also a recipe for undermining the seed sovereignty and food security of small farmers and agrarian economies.

The Gates Foundation is the new World Bank and the New Monsanto when it comes to using finance to control seeds and agriculture and influence policies. It is a major funder of the CGIAR system – and through its funding, it is accelerating the transfer of research and seeds to corporations, facilitating intellectual property piracy and the creation of seed monopolies through intellectual property laws and seed regulation. Since 2003, CGIAR centres have received more than $720 million from Bill Gates. Farmers' seed heritage is held in the seed banks of CGIAR centres. By taking control of the CGIAR system, Gates is attempting to centrally control all agriculture research, as well as the world food system. He is working towards 'One CGIAR', which is part of his 'One Agriculture' or the Gates Ag One programme that aims to provide small-holder farmers in developing countries, many of whom are women, with access to 'affordable' tools and innovations to improve crop productivity and adapt to the effects of climate change. The company intends to enable the advancement of resilient, yield-enhancing seeds and traits globally and facilitate the introduction of those breakthroughs into specific crops essential to small-holder farmers, particularly in sub-Saharan Africa, home to around one billion people, and South Asia, with a population of about 1.8 billion. In these regions, approximately 60 per cent of the population lives in rural areas that typically depend on small-holder agriculture for food and income. According to its mission statement, the Gates Foundation works to help all people lead healthy, productive lives.

Gates does not create climate traits, he pirates them. By taking control of the CGIAR systems, he controls the world's climate-resilient seeds. At a time when the world is waking up to the value of decentralisation, localisation and food democracy, Gates wants a centralised system of control that will destroy the biodiversity of crops, of ecosystems and farming systems adapted to climates, and the diversity of food cultures. Gates has proposed a 'One CGIAR Common Board', and an annual budget increase from $850 million to $2 billion, in an attempt to control the research of 1,500 scientists and the 773,000 seed accessions collected from farmers. The dystopian vision of 'One Agriculture' for a world of diversity is intolerant of diversity, democracy and sovereignty, extinguishing seed freedom, knowledge freedom, food freedom and health freedom.

Bill Gates (along with the Rockefeller Foundation) is also investing heavily in collecting seeds from across the world and storing them in a facility in Svalbard in the Arctic – the 'doomsday vault'. The Svalbard Global Seed Vault (Norwegian: *Svalbard globale frøhvelv*) in the Arctic Svalbard archipelago holds the global collection of seeds and is run in association with the CGIAR centres and the Crop Trust. In May 2014, a total of 820,000 accessions were stored in the Vault, a mere fraction of its 4.5 million accessions capacity.

The largest numbers of accessions stored there are varieties of rice, wheat and barley: more than 150,000 samples of wheat and rice and close to 80,000 samples of barley. Other well represented crops are sorghum (>50,000 accessions), Phaseolus bean species (>40,000), maize (>35,000), cowpea (>30,000), soybean (>25,000), and kikuyu grass and chickpea, both with more than 20,000 seed samples. Crops such as potatoes, peanuts, Cajanus beans, oats and rye, alfalfa, the cereal hybrid, Tritikosecale and Brassicas are represented by between 10,000 and 20,000 seed samples. The Crop Trust, based in Germany, funds and coordinates the Svalbard Seed Vault; among its funders are the Gates Foundation and Crop Life, Dupont/ Pioneer Hi Bred, KWS SAAT AG, and Syngenta AG.

One major project that Gates supports is Stress Tolerant Rice for Africa and South Asia (STRASA). From 2007 to 2010, and 2011 to 2014, he gave US $20 million to IRRI for each phase of the project. The third phase, from 2014 to 2019, had a budget of US $32.77 million. Gates claims that he suborned the innovation of the flood-tolerant rice variety, Swarna-Sub1, but its flood-tolerant traits were derived from an Indian variety, Swarna, and bred into Sub 1 rice using marker-assisted selection.

What corporations and the Gates Foundation are doing is obtaining farmers' varieties with known climate-resilient traits from public gene banks, mapping their genome, and taking out patents on the basis of guesswork and speculation about which part of the genome contributes to the known trait. Patenting life through biopiracy or genetic engineering is rapidly giving way to patenting life through mapping the genome. Navdanya's Community Seed Bank in Orissa has conserved more than 800 rice varieties and multiplied and distributed salt-tolerant and flood-tolerant varieties. The 'innovation' to evolve these climate-resilient traits has occurred cumulatively and collectively over thousands of years. These traits and crops are a commons, but are now being presented as the 'invention' of 'scientists' who rename the flood-tolerant property in the farmer's variety such as 'Dhullaputia' from Orissa as the Sub1A or the submergence tolerant gene. Using marker-assisted selection (not transgenics), the researchers were able to isolate the submergence tolerant gene, Sub1A, and then transfer it to a rice variety that is grown on more than five million hectares in India and Bangladesh, known as Swarna. Most rice can tolerate flooding for only a few days, but researchers say the Swarna-Sub1 can withstand submergence for two weeks without affecting yields.

This is a scientifically flawed claim based on genetic reductionism, because flood tolerance, like other climate-resilient traits such as salt tolerance and drought tolerance, are multigenetic traits; they cannot be identified as a 'Sub1A gene'. Because it is not 'a gene', it has been referred to as 'Submergence tolerance 1 (Sub1) Quantitative trait locus (QTL)'.

Instead of Gates controlling climate-resilient seeds through his seed empire, we need to propagate seeds of reliance, seeds of justice and seeds of freedom through community seed banks everywhere. Reclaiming seed freedom is reclaiming the commons of seed. Today, biopiracy and the enclosure of biodiversity and indigenous knowledge are carried out through the convergence of IT and biotechnology.

A map does not create either the land or life on it, but it becomes an instrument of ownership and control. Similarly, a genomic map does not create a living seed that grows into a plant, nourishes the soil and the web of life, and gives us food. Without knowing the seed, without breeding, by just using the passport data of seeds in gene banks, billionaires like Gates and the Poison Cartel they collaborate with are using Artificial Intelligence (AI) and algorithms to guess which part of the genome contributes to a particular trait that farmers have bred and evolved with deep knowledge and co-creativity. One such platform is called ATHLETE.

Evogene Ltd. has patented a computer programme for reading the genome. Its proprietary in silico 'gene discovery technology' is called ATHLETE. (In silico, as opposed to in vivo or in vitro, refers to investigations performed through the use of a computer or computer simulation.)

ATHLETE is the company's proprietary computer database and analysis programme for finding gene function by comparing sequences from as many different plant species, tissues, organs, and growth conditions as possible. Evogene says its database consists of eight million expressed sequences, 400,000 'proprietary gene clusters', and 30 plant species. The programme clusters sequences according to a variety of criteria, and then determines which gene candidates to investigate further. It is an informed winnowing process.

Over the last two decades, humanity has taken action and instituted laws to protect the biodiversity of the planet and the rights of farmers to seed and the rights of consumers to safety. These laws include: The Convention on Biological Diversity (CBD); the Cartagena Protocol on Biosafety to the CBD; and the International Treaty on Plant Genetic Resources for Food and Agriculture.

José Esquinas-Alcazar, the architect of the Seed Treaty at the FAO, which recognises and protects farmers' rights and seed sovereignty, warns how the treaty to protect agricultural biodiversity is being undermined by the Gates-led Digital Genomic Mapping of ex situ collections.

Aidé Jiménez-Martínez[18] and Adelita San Vicente Tello[19] show how the Nagoya Protocol (under the CBD negotiated by the world community to protect biodiversity and seed sovereignty) could be rendered meaningless unless digital sequence information is included in its ambit. Bypassing international treaties to protect seed sovereignty is the clear aim of the digital colonisation of seed. The International Digital Council for Food and Agriculture, presented at the Global Forum for Food and Agriculture in Berlin in 2019, proposed data gathering on a global scale:

> These practices would extend to genetic digital sequence information (DSI or digital DNA) and could pre-empt already-contentious access and benefit sharing (ABS) negotiations in the Biodiversity Convention and the FAO Seed Treaty.

This is why digital colonisers say that data is the new gold.

There are over seven billion people on the planet today, a figure that is expected to reach 9.6 billion by 2050. Overall food production needs to double in a relatively short period of time in order to meet the food demand of the world's population. It is, therefore, imperative for different agricultural stakeholders to revolutionise traditional agriculture.

Reclaiming the seed is the first step in reclaiming our food and health freedom, a vital step in defending freedom and democracy. In a time of seed imperialism and food dictatorship, reclaiming food by reclaiming the seed becomes a most revolutionary act.

*****

According to the FAO, one billion people are permanently hungry and more than two billion suffer from food-related diseases. Contrary to the fallacy that small farmers and their agroecological systems are unproductive and, therefore, dispensable, small farmers in fact provide for 70 per cent of global food requirement using just 30 per cent of the resources that go into agriculture. In direct contrast, industrial agriculture uses 70 per cent of the resources while providing only 30 per cent of our food. Additionally, it is creating a health crisis by producing nutritionally empty toxic commodities, contributing 75 per cent of food-related chronic diseases.

Organic farming takes the excess carbon dioxide from the atmosphere, where it doesn't belong, and through photosynthesis returns it to the soil, where it does. It also increases the soil's water-retaining capacity, contributing to resilience in times of drought, flood and other climate extremes. Chemical agriculture, on the other hand, does not return organic matter and fertility to the soil, so essential for maintaining nature's life cycle. Instead, it contributes to the desertification and degradation of land by destroying its natural water-retention capacity, and the use of toxins and poisons are killing bees, butterflies, insects and birds, paving the way to the sixth mass extinction.

Nonetheless, we are sowing the seeds of another future. Across the globe, small farmers and gardeners are already implementing biodiversity-based, chemical-free, organic agriculture; they are practising agroecology and preserving and developing their soils and seeds. They are feeding their communities healthy, nutritious food even as they rejuvenate the soil and the planet. They are sowing the seeds of food democracy – a food system that is in the hands of farmers and consumers, devoid of corporate control, poisons, food miles, and plastics; a food system that nourishes the planet and all humans.

We cannot address climate change – and its very real consequences – without recognising the central role of the industrial and globalised food system, which contributes to greenhouse gas emissions through deforestation; concentrated animal feeding operations (CAFOs); plastic and aluminium packaging; long-distance transport; and food waste. We cannot arrest climate change without small-scale, ecological agriculture. What we eat, how we grow the food we eat, and how we distribute it will determine whether humanity survives or pushes itself, and other species, to extinction.

Food is not a commodity; it is not 'stuff' put together mechanically and artificially in labs and factories. Industrial food systems have reduced food to a commodity, to some 'thing' that can be created in a lab. But food is life. It is a product of the contributions of all beings that make up the food web, and it holds the potential for maintaining and regenerating the web of life. Food also holds the potential for health and disease, depending on how it is grown and processed. It is, therefore, the living currency of the web of life.

Despite the documented failures of industrial agriculture, it has continued to reinvent and reinvest in its future based on 'fake farming and fake food', first with chemical fertilisers, then with GMOs, and more recently with Big Data, derived from surveillance drones and spyware. Fake food, including fake meat, is a product of the fake farming that produces fake food commodities. Referring to plant-based meat companies, Bob Reiter, Bayer's Head of Research and Development, has said, 'They are sourcing different types of crops, and that also could create opportunity for us, being a company that is a plant-breeding company.' Clearly, even though more and more people are shifting to agroecology and organic food as cultivators and consumers, and more and more communities are creating local, ecological systems based on diversity, the Poison Cartel is still hoping that fake food will create new markets for fake farming. It is trying hard to push us further down the dead-end path by making farming robotic and mechanistic, increasingly disconnected from the living intelligence of the earth. In 2018, Bayer-Monsanto claimed that it was implementing a programme in the American midwest to deliver IntelliScanSM field guides and IntelliSeedSM custom planting recommendations to farmers. This would be the first phase of Monsanto Prescriptive Ag Solutions, a programme 'with the vision of providing growers increased confidence in seed choice and the best placement and plant populations for their farm'.

But data is not knowledge. It does not provide insights into how the solution to climate change lies in the soil, nor how the rich soil–food web is composed of bacteria, fungi and earthworms essential for soil fertility. It is just another commodity to make the farmer less connected to the earth, outsourcing his or her mind to Big Agribusiness.

The EAT Forum was set up in 2016 in Stockholm, and I was invited to its first meeting. Very rapidly it started to drive the global food and nutrition agenda. Along with *The Lancet*, a leading medical journal, it set up the EAT-Lancet Commission on Food, Planet, Health. Many of my friends were on it, including Dr Srinath Reddy of the Public Health Foundation of India (PHFI) and Prof. Tim Lang,[20] who had worked with me in the IFG and wrote a book, *The New Protectionism: Protecting the Future against Free Trade*, with another IFG colleague, Colin Hines. I respect them deeply and have worked with them closely over the years. However, the EAT-Lancet Report, released in 2019, did not reflect the thinking of the leading health and nutrition experts of our times, who recognise that food has to be local and diverse to be healthy. The Report has an overall thrust towards a global corporate food agenda, based on more chemical and fertiliser use, more industrial processing of food, and one planetary diet. It has tried to impose a monoculture diet of chemically-grown, hyper-industrially processed food on the world, claiming that a 'healthy and sustainable (plant–based) diet' protects the health of the planet and of people, completely ignoring the widespread chronic disease epidemic that is related to the pesticides and toxins used in the cultivation and processing of food produced by industrial agriculture.

To produce this report, EAT Forum partnered with FRESH (of the junk food industry) and Big Ag, including Bayer, BASF, Cargill, Yara, and PepsiCo, among others. (Bayer is the biggest GMO seed and agrochemical company after its merger with Monsanto, while Yara is the biggest chemical fertiliser corporation in the world.) Thus, the report promotes what can also be called 'the Poison Cartel diet', where *real* health and sustainability are concepts alien to its authors and promoters.

FRESH, together with the Poison Cartel, has contributed to 50 per cent of greenhouse gases, and to 75 per cent of the chronic disease epidemic related to chemicals in food, loss of diversity in the diet, industrially-processed and junk food, and fake food. From feeding the world with fake farming, the rhetoric has shifted to saving the planet and people's health through fake food. In other words, the EAT Report is more of an advertisement for industrial and fake food, using the fig-leaf of 'plant-based diet'. Instead of recognising the role of organic farming and agroecology in providing sustainable ways of repairing the broken nitrogen cycle, the EAT Report recommends the 'redistribution of global use of nitrogen and phosphorus' which, in effect, says that chemicals should continue to be widespread in the countries of the South. This is precisely what the Gates Foundation/AGRA is doing.

Six months before the release of the EAT Report, Navdanya International published the *Food for Health Manifesto*, written by noted health experts and ecologists, which identified toxins as the leading cause of the disease epidemic. Toxins are the elephant in the planetary and human health room that the EAT Report is completely silent on. More than half a century after Rachel Carson's *Silent Spring*; more than thirty-four years after the Bhopal genocide; one year after the UN Special Rapporteur on the Right to Food put out his report on pesticides; and a few months after the Johnson trial firmly established that Round-Up is a carcinogen, the EAT Report failed to mention that toxins are driving species to extinction and have led to an epidemic of cancers, neurological problems, endocrine disruption, and infertility.

In his article 'How our Commitment to Consumers and Our Planet Led Us to Use GM Soy', Pat Brown, CEO and Founder of Impossible Foods, writes: 'We sought the safest and most environmentally responsible option that would allow us to scale our production and provide the Impossible Burger to consumers at a reasonable cost.' This, despite the fact that Round-Up-sprayed GMO soya has already caused massive ecological devastation as well as serious chronic health problems worldwide.

The Impossible Foods' Impossible Burger cannot be considered a 'safe' option, both because of its high levels of glyphosate and its effect on our gut microbiome. According to Zen Honeycutt of Moms Across America, 'The levels of glyphosate detected in the Impossible Burger by Health Research Institute Laboratories were 11 times higher than the Beyond Meat Burger.' Furthermore, Round-Up residues disrupt the shikimate pathway in our body, the all-important biological pathway in our gut bacteria, on which we depend for the supply of essential nutrients that are deficient in our diet.

The Poison Cartel, Big Food and Big Money are investing millions in the fake food industry to support the mass proliferation of 'fake' eggs, dairy and meat. Indeed, the promotion of fake foods seems to have something to do with giving new life to the failing GMO agriculture and junk food industry, and the threat from the rising consciousness and awareness that organic, local, fresh food is real food. Consequently, investment in 'plant-based food companies' has soared from nearly nil in 2009 to $600 million in 2018 – and these companies are looking for more.

Among the new players pushing the fake food agenda are companies like Beyond Meat (BYND.O) and Impossible Foods, and even traditional meat producers like Tyson Foods Inc. (TSN.N), Maple Leaf Foods Inc. (MFI.TO) and Perdue Farms. According to Pat Brown of Impossible Foods, 'If there's one thing we know, it's that when an ancient, unimprovable technology counters a better technology that is continuously improvable, it's just a matter of time before the game is over…. I think our investors see this as a $3 trillion opportunity.'

This is about profits and control. Brown, and those gambling on the Fake Food Goldrush, have no discernible knowledge or compassion for living beings and the web of life. Their sudden awakening to plant-based diets, including GMO soya, is an ontological violation of food as a living system that connects us to the ecosystem and other beings. Ecological sciences are based on the recognition of the interconnections and interrelatedness between humans and nature, between diverse organisms, and within all living systems, including the human body. They have thus evolved as an ecological and systems science, not a fragmented and reductionist one. Diets have evolved according to climates and the local biodiversity that the climate supports. The biodiversity of the soil, of the plants and our gut microbiome is one continuum.

Fake food is building on a century-and-a-half of food imperialism and food colonisation; it is another phase in the history of food imperialism, dismissive of the knowledge and cultures it has colonised. While farmers in India have always known that pulses fix nitrogen, the West was industrialising agriculture, based on synthetic nitrogen. While we consumed a variety of lentils, as part of our daily 'dal-roti' meals, the British, who had no idea about the rich nutritional value of pulses, reduced them to animal food – *chana* became chickpea, *gahat* became horse gram, *toor* became pigeon-pea.

When GMO soya oil started being dumped in India and our local oils and cold press units in villages were made illegal, women from India's poorest communities mobilised to bring back the mustard. When our mustard oil was banned in 1998, women from the slums of Delhi reached out to me for help to 'bring our mustard back'. I asked them to give me a week to understand the situation and realised that the GMO soya lobby wanted to capture the Indian edible oil market. I wrote a small booklet, *Mustard vs. Soya*, on how our diverse oilseeds were healthier and produced more oil than soya. We also created the National Alliance of Women for Food Sovereignty, or Mahila Anna Swaraj, to defend our food sovereignty. We planned a Sarson (Mustard) Satyagraha in Connaught Place, in the heart of New Delhi; Kamla Bhasin, Maya Goburdhun,[21] and others led the protest. We met the then President of India, K.R. Narayanan and his wife, and at Dilli Haat, another venue in the city, we extracted mustard oil with a small cold press mill.

Soya is a gift of East Asia, where it has been a food for millennia; it was only eaten as fermented food to rid it of its anti-nutritive factors. Today, GMO soya has created soya imperialism. It continues the destruction of the diversity of rich edible oils and plant-based proteins. For example, India, the land of rich diversity in edible oils – mustard, sesame, coconut, linseed, groundnut, etc. – is now 70 per cent dependent on imports of palm oil and GM soya oil.

But what goes around comes around. Our artisanal processed coconut and mustard oils are being recognised as healthy, in spite of the decades-old, pseudo-scientific propaganda against them by the industrial food processing lobby that has been promoting trans-fats in the diet by influencing the food policy, trade policy, scientific research, and the huge money they spend on misinformation and misleading advertising.

The organic movement has grown steadily over the past four decades across India and the world. Howard's *An Agricultural Testament* inspired Eve Balfour to start the Soil Association to promote organic farming in the UK way back in the late 1940s. In the US, J.I. Rodale started the organic movement with the Rodale Institute around the same time. Later, the two organisations joined hands to create the global organic movement, IFOAM, a worldwide umbrella organisation for the organic agriculture movement, representing close to 800 affiliates in 117 countries.

Over the years, I have worked with seven states in India to grow organic – Uttarakhand, Kerala, Madhya Pradesh, Orissa, Himachal Pradesh, Ladakh and Sikkim. Pawan Chamling, former Chief Minister of Sikkim, played an instrumental role in turning Sikkim into a fully organic state in 2017. When our neighbour, Bhutan – which has adopted the concept of Gross National Happiness (GNH) as opposed to Gross Domestic Product (GDP) as a measure of national prosperity – decided to embrace organic farming, their Prime Minster, Jigme Thinley, invited me to help Bhutan make the change. As he wrote in his letter to me in 2009: 'I cannot see growing happiness without growing organic.' In 2013, organic movements of the world joined hands to launch Organic Himalaya in Bhutan, a movement we continue to build and strengthen. We issued the Thimpu Declaration for an Organic Himalaya on the occasion and further developed the idea at a conference on biodiversity in Sikkim in 2018. The Indian states of Himachal Pradesh, Ladakh, Uttarakhand and Sikkim, as well as Nepal and Bhutan, are part of Organic Himalaya.

Today, we stand at the precipice of a planetary emergency, a health emergency, and a crisis of farmers' livelihoods. Fake food will accelerate the collapse by promoting and sustaining the failed fake farming model of industrial agriculture. Real food gives us a chance to rejuvenate the earth, our food economies, food sovereignty and food cultures through real farming and agroecology.

Notes

[1] Political psychologist, social theorist and critic.

[2] Academic, author, former Director of the Centre for the Study of Developing Societies and founding member of the Institute of Chinese Studies.

[3] Social scientist who co-founded Lokayan and Lokniti.

[4] A noted environment educator, Sarabhai is the founder and director of the Centre for Environment Education. He was conferred the Padma Shri in 2012.

[5] Late Director of the Department of Agriculture, Orissa, he gave up his job to return to his native Hoshangabad, Madhya Pradesh, to start a Gandhian ashram, Gram Sewa Samiti.

[6] An alumnus of the Indian Institute of Management (IIM) Calcutta, Mathen was vice president, rural development, Arvind Mills, Ahmedabad.

[7] Gandhian and social activist associated with Sevagram Ashram Pratishthan in Wardha, Maharashtra.

[8] A prominent Indian farm leader and former president of the Bharatiya Kisan Union, he played a key role in the farmers' protests against the new farm laws in 2021.

[9] A French molecular biologist, political advisor and activist on genetically modified organisms and foods.

[10] Historian, writer and essayist.

[11] A geneticist, plant breeder and innovator, Ceccarelli has spent over five decades researching agriculture for development to aid poor farming communities.

[12] Geier has over 40 years' experience in the field of agricultural and environmental politics. He served as director of the International Federation of Agriculture Movements (IFOAM) for 18 years and has authored numerous books on organic agriculture.

[13] Barker is associated with the International Forum on Globalisation, a think-tank that analyses and critiques forms of economic globalisation, and is a member of the Committee on the Future of Food and Agriculture commissioned by the government of Tuscany.

[14] A Chilean-born agronomist and entomologist, he is Professor of Agroecology at the University of California, Berkeley.

[15] A microbial ecologist and mycologist at the University of California, Berkeley, he is a vocal critic of biotechnology.

[16] A Welsh writer, Dr John has worked as a field scientist in Antarctica, has travelled widely in the Arctic, Antarctic and Scandinavia, and is actively involved in environmental and community organisations.

[17] The Iraq-born and raised Al-Natheema is a poet, author and translator. She is the founder of the Institute of Near Eastern & African Studies (INEAS), a US-based non-profit organisation.

[18] Jiménez-Martínez is Director of Regulations of Biosafety, Biodiversity and Genetic Resources, SEMARNAT, Mexico.

[19] Tello is Director General of the Primary Sector and Natural Resources, SEMARNAT, Mexico.

[20] Emeritus professor of food policy at City, University of London's Centre for Food Policy since 2002; Lang also set up the Cambridge Forum for Sustainability and the Environment.

[21] Director, Navdanya.

Diverse
Women for
Diversity

MY INVOLVEMENT WITH CHIPKO AND WITH MANY ECOLOGICAL movements since then has convinced me that women are in the vanguard for the defence of the earth. At the 1985 UN End of Decade Conference on Women, I gave a talk on how the roots of the domination and exploitation of nature and of women are the same, and that the liberation of both are also the same, since we are a part of nature, not her masters or owners. The late Nobel Laureate, Wangari Maathai, also spoke at the Women and Environment session. Ritu Menon of the feminist press Kali for Women heard me and insisted I write a book on the interconnectedness of women's rights and the rights of nature. I resisted initially, having given up academic research and not wanting to be a part of the publishing rat race. Ritu persuaded me by saying that writing, too, can be a subversive activity. She published *Staying Alive* in South Asia, while Zed Books published it in the UK. Zed was also the worldwide publisher of Maria Mies's book, *Patriarchy and Accumulation on a World Scale*. I knew Maria well and always stayed with her when I visited Cologne. Both our books became bestsellers, so Zed asked us to write a book together on ecofeminism. Neither of us coined that term; it was first created by the French feminist, Françoise d'Eaubonne, in her 1974 book, *Le Féminisme ou la Mort* (Feminism or Death: How the Women's Movement Can Save the Planet).

For me, the convergence of nature's creative power and women's power in creative nonviolent resistance against the destruction of nature became apparent during my participation in the Chipko movement. I wrote about it in *Staying Alive*, and Maria and I developed it together in our book *Ecofeminism*. The book grew organically, like our thinking and practice. Maria was in Cologne, while I was in Dehradun, India. When the International Centre for Integrated Mountain Development (ICIMOD) invited me to set up their women's studies centre I used the opportunity to organise a conference, and invited Maria to give the keynote address. We had agreed to write our chapters independently, responding to events and processes as they unfolded. When we met, we found our ideas and involvement in movements had evolved in resonance because we were both responding with the awareness that nature is alive and has rights, and that women are at the heart of the creative and productive endeavours of humanity.

Maria and I have walked together over many years. At the Leipzig Conference on Plant Genetic Resources we drafted and issued The Leipzig Appeal for Women's Food Security on June 20, 1996. At the 1996 World Food Summit in Rome, the Leipzig Appeal became a declaration, 'Keeping Food Security in Women's Hands', endorsed by more than 100,000 women. This declaration launched the global women's food sovereignty movement.

Maria was also part of the global movement Diverse Women for Diversity, of which I was a member. It was co-founded by Jean Grossholtz[1] and Beth Burrows[2] of the US, and Christine Von Weizsäcker[3] of Germany. Others who became part of this movement were Wangari Maathai and Maria Zuniga.[4] We were the women defending biodiversity in the face of the rising threat of GMOs, patents, and monocultures during the negotiations on the Convention on Biological Diversity (CBD) and the WTO meetings, including the WTO Ministerial Conference in Seattle in 1999. We launched Diverse Women for Diversity at the CBD meeting in Bratislava in 1998.

Maria is 90 years old now. The last time we met was when I was invited to Bonn by the movement against Monsanto, now Bayer. Maria joined me on stage and sang a song she had composed against genetic engineering in the 1980s.

In India, the Diverse Women for Diversity movement evolved as Mahila Anna Swaraj (Women's Food Sovereignty). We created a national alliance of women's groups and engaged in a satyagraha against the banning of mustard and the dumping of GMO soya oil. With women's groups like Kudumbashree and SEWA (Self-Employed Women's Association), Navdanya members continue to build a movement for diversity, against monocultures of the mind; for localisation of the food economy, against globalisation; and for economies of caring and sharing.

Forests and farming are systems where the structures and paradigms of capitalist patriarchy are very clear and evident. Women everywhere are contesting monocultures, chemicals, and commodification. Industrial agriculture, seed monopolies and seed patents are based on a larger vision of capitalist patriarchy, which is born of the illusion that nature is dead or inert matter, and that women are passive – they have bodies without minds. As Descartes said, powerful men are 'thinking things without bodies'. Having defined nature as dead and women as non-creative and unproductive, capitalist patriarchy appropriates what nature and women create and produce and present it as created by a dead construct, 'capital'.

Mechanical philosophy, the rise of colonialism, the construction of nature as inert matter to be exploited as raw material, and the violence against women as witches was a contemporaneous process. Nine million people, mostly women, were burnt as witches in Europe. The same worldview and structures that objectify nature and dominate the earth also objectify women.

Agroecology and ecofeminism dispel these illusions, recognise the creativity of nature, of women, and of farmers, and create knowledge systems and economic structures that work according to ecological laws of nature, and advocate social and gender justice based on the principle of all humans being equal.

In 1991, feminist-activist Kamla Bhasin, who was with the FAO, asked me to write a report on women farmers of India. It was published under the title 'Most Farmers of India are Women'. Kamla and I shared many aspects of our personal and political lives. Her daughter, Meeto, and my son, Kartikey, went to the same school and cared for each other like brother and sister. In 2021, Kamla was diagnosed with liver cancer and passed away on September 25 of the same year. India lost one of her most inspiring feminists; I lost a dear friend and a fellow traveller.

Ecofeminism as a philosophy is based on interconnectedness. It allows us to see how the invasion of forests by agribusiness has contributed to the emergence of new infectious diseases like SARS, Ebola, Zika, and HIV. Industrial food systems are also leading to chronic disease pandemics. The Poison Cartel that made the chemicals that killed people in concentration camps and during WWII continues to sell these chemicals as agrichemicals, causing a cancer pandemic. And the same Poison Cartel is Big Pharma that has patents on cancer drugs and COVID vaccines.

Ecofeminism allows us to see the interconnections and also find solutions that lie in our hands. This is where agroecology and healthy eating for immunity become relevant. And as Big Tech and Silicon Valley join hands with the Poison Cartel to impose patented lab-made fake food, our ecological alternatives become even more important for our lives and our freedoms.

*****

Women make the most significant contribution to food security. They produce more than half of the world's food and provide more than 80 per cent of the food needs of food insecure households and regions. Food security is, therefore, directly linked to women's food-producing capacity. Constraints on women's capacity to do so then lead to the erosion of food security, especially for impoverished households in poor regions. From the field to the kitchen, from seed to food, women's strength is diversity, and their capacities are diminished when this diversity is destroyed. Diversity is the pattern of women's work; of the pattern of women's planting and sowing of food crops; and the pattern of women's food processing.

The dominant systems of economics have rendered women's work as food providers invisible because they provide for the household and perform multiple tasks involving diverse skills. They have remained invisible as farmers in spite of their contribution to farming, and economists tend not to record their production as 'work' or as 'production boundary'. This problem of data collection arises not because too few women work but because too many women do too much work. There is a conceptual inability on the part of statisticians and researchers to define women's work inside the house and outside (and farming is usually part of both). It is also related to the fact that although women work to sustain their families and communities, most of their work is not measured in wages. It is invisible because women are kept out of market-related or remunerative work.

Science and technology, too, have rendered women's knowledge and productivity invisible by ignoring the dimension of diversity in agricultural production. As the FAO report *Women Feed the World* reveals, women use more plant diversity, both cultivated and uncultivated, than agricultural scientists know about. In Nigerian home gardens, women plant 18-57 plant species, while in the countries of sub-Saharan Africa women cultivate as many as 120 different plants alongside the cash crops that are managed by men. In Guatemala, home gardens of less than 0.1 ha have more than 10 tree and crop species, and in Thailand, research has found up to 230 plant species in home gardens. In India, women use 150 different species of plants for vegetables, fodder and healthcare, while in the east Indian state of West Bengal, 124 species of 'weed' collected from rice fields have been identified as having economic value. Clearly, women manage and produce diversity even as the dominant paradigm of agriculture promotes the perception that monocultures produce more. Monocultures do not actually produce more, they just control more. According to an FAO World Food Day report, a study in eastern Nigeria found that home gardens, occupying only two per cent of a household's farmland, accounted for half the farm's total output. Women's agriculture technologies are thus 49 times more productive than conventional agriculture. So, if women's knowledge was not being rendered invisible, the use of the two per cent land under polyculture systems would be the path to ensuring food security. Instead, these highly productive systems are being destroyed in the name of producing 'more' food. Hazardous genetic engineering technologies are being introduced on the grounds that they increase land productivity fourfold. Even if this were true, it would still be ten times less than women's way of increasing productivity.

A study in Francesca Bray's 'Agriculture for Developing Nations' (*Scientific American*, 1994) comparing traditional polycultures with industrial monocultures shows that a polyculture system can produce 100 units of food from five units of input, whereas an industrial system requires 300 units of input to produce the same quantity. The polyculture system is thus 60 times more productive than a monoculture system. The 295 units of wasted input can provide 5,900 units of additional food. Productivity in traditional farming practices has always been high as very few external inputs are required.

Just as women's ways of growing food produce more even as they conserve resources, women's ways of food processing conserve more nutrition, too. Hand-pounding rice or milling rice with a foot-operated mortar and pestle preserves more protein, fat, fibre and minerals in it. So, when mechanical hullers replace hand-pounding – as in Bangladesh, where 700 new mills supplanted the paid work of 100,000 to 140,000 women in one year by reducing the labour input from 270 hours per ton to five – they not only rob women of work and livelihoods, they also rob consumers of essential nutrients. And yet, patriarchal economies insist that this process of destroying the value of food is in fact 'value addition'.

Feeding the world requires producing more food with fewer resources, i.e., producing more with less. In this, women are experts, and their expertise needs to filter into our institutions of agricultural research and development. However, instead of building on women's expertise in feeding the world through diversity, the dominant system is rushing headlong towards destroying diversity and women's food-producing capacities.

The lack of property rights for women is a major constraint when it comes to women's capacity to feed the world. These rights include the right to land, as well as the right to common resources like water and biodiversity. Women have traditionally been the custodians of biodiversity, but new Intellectual Property Rights (IPRs) are alienating their rights to biodiversity and erasing their innovation, embodied in agricultural biodiversity. If the destruction of women's capacity to feed the world is to be stopped, IPRs need to evolve sui generis systems that recognise and protect their collective and informal innovation. The Agreement on Trade-Related Aspects of Intellectual Property Rights (TRIPS) has to ensure that women's rights to their knowledge and biodiversity are protected.

While women are being denied their rights and resources and we are seeing the feminisation of subsistence agriculture, dominant agriculture is showing increasing signs of masculinisation as it appropriates resources and rights from women in subsistence agriculture and presents itself as the only alternative for feeding the world.

Women farmers in the global South are predominantly small farmers. They provide the basis of food security, and they provide food security in partnership with other species. This partnership between woman and biodiversity has kept the world fed throughout history, at present, and will feed the world in the future as well. It is this partnership that needs to be preserved and promoted to ensure food security.

*****

A global pandemic has shut down economies and societies. Chronic diseases, including cancer, diabetes and infertility are increasing as real food is substituted with unhealthy, industrial ultra-processed fare. Fires are burning in Serbia, in the Amazon and in the US. Hurricanes are striking with increased intensity and frequency in Korea and India. Floods have submerged villages in Sudan and Australia, and droughts are drying up the rainforests, which should be creating and receiving heavy rain. I call this Climate Chaos. It is a result of disrupting the ecological cycles and the biodiversity of species that maintain the living planet and the infrastructure of life. Non-sustainability and multiple emergencies are emerging as a threat to the very survival of the human species. Non-sustainability is rooted in separation and the invisibility of women's knowledge and work. The health emergency that the corona virus has created is connected to the emergency of extinction and the disappearance of species, and it is also connected to the climate emergency.

Indeed, all emergencies are rooted in the industrial, mechanistic, militaristic, anthropocentric, patriarchal worldview of humans as separate from nature, and superior to other beings who can be owned, manipulated and controlled for profit and power; of men as superior to women; of whites as superior to the diverse colours of humanity. The global interconnected crises are also rooted in an economic model based on extractivism, the illusion of endless growth and limitless greed, which systematically violates the integrity of species and the limits of the ecosystem.

Women's knowledge was made invisible with the rise of Francis Bacon's *The Masculine Birth of Time*, which displaced ecological and embodied knowledge as non-knowledge, and elevated mechanistic reductionist knowledge, based on the denial of life and its interconnections, as 'science'.

In agriculture, the science of life is called agroecology. It is a systems science based on nature's ecological laws. It is the umbrella term for all ecological agriculture systems that work with nature, for life. These include permaculture, biodynamic agriculture, natural farming and organic farming. Rural women have kept systems of agroecology alive and provide the link between human societies and the earth. They become the bridge to the future by embodying the time-tested principles of sustainability in their everyday practices.

There are two paradigms of science and knowledge. The first paradigm is holistic and aligned with nature and her ecological laws. This is the paradigm that rural women have evolved, sustained and renewed over thousands of years of sustainable agriculture policy. It is a dynamic paradigm that evolves with time to respond to changing social and environmental conditions. This is the paradigm for the 21st century.

The second paradigm is reductionist, mechanistic and extractive. Globalisation and free trade rules, shaped by corporate power and capitalist patriarchy, work towards the marginalisation of women, their knowledge, and their economic and political power.

My research over the past three-and-a-half decades has shown that biodiversity is the path to take in order to feed the world. Monocultures produce more yields of rice, wheat, corn, soya, but these commodities are rotting in godowns, being fed to animals or being converted to biofuel to drive cars. We need to move away from measuring 'yield per acre' of nutritionally empty monocultures, produced at high cost, to measuring the 'nutrition per acre' of crop diversity. In a biodiversity paradigm, the real metric is 'nutrition per acre' or 'health per acre', not 'yield per acre'. To evolve alternatives to the misleading construct of 'yield', I started to assess the biodiversity based productivity of farms, based on the total biodiverse output of farming systems, and comparing the output of biodiverse organic farms with chemical monocultures. On the basis of biodiversity based productivity, Navdanya has evolved the metric of biodiversity and 'nutrition per acre' using the nutritional data of the National Institute of Nutrition on indigenous foods.

I questioned the false metric of 'yield per acre', which merely measures the weight of a commodity that leaves the farm, and not the health of the farm, the farmer, or the food. It only measures the yield of a single commodity, excluding the high cost of external inputs and their ecological and social impact. The false claim that chemicals produce more food and are necessary to feed the world is based on the claim of productivity gains and the higher 'yield per acre' of the Green Revolution, but this is a false comparison.

Rural women do not think and work through commodification and reductionism. They manage the farm as their home, have a sense of place and belonging, and develop systems based on the health of the soil along with the health of their families and communities. As the Navdanya study *Health Per Acre* shows, if, instead of the chemical intensification and capital intensification of our agriculture, we intensified food production ecologically and in harmony with biodiversity, we could feed twice the population of India with healthy, balanced food. And this principle of biodiversity intensification applies to every ecosystem. Rural women grow health by growing biodiversity – in the soil, in our fields, and in our diet. It is this science of biodiversity and interconnectedness between living soil, living seed, living food, living economies of well-being that I have been researching and practising and promoting through Navdanya. This rich biodiversity of living knowledge is conserved and regenerated in every culture by rural women.

Women's seed sovereignty is central to their knowledge and economic sovereignty; they have been the guardians of seeds for centuries. In a study by Navdanya, it was found that in 60 per cent of the examples cited in the study, it was the women who decided which type of seed to use. On farms where they use their own seeds, the decision is exclusively made by women 81.2 per cent of the time. At Navdanya, we have reclaimed the seed as commons and created 150 community seed banks to protect the biodiversity of the seed. Across the world, we have also inspired the seed freedom movement that has created a new consciousness about seed sovereignty.

We have protected 4,000 varieties of rice, 250 varieties of wheat, 11 varieties of barley, 5 barnyard millet varieties, 10 varieties of oats, 6 varieties of finger millet, 3 varieties of foxtail millet, 7 varieties of mustard, and 450 different medicinal and perennial and deciduous trees used for wood.

The seeds that we have conserved have aided in recovery after cyclones, tsunamis and recurring floods. Navdanya has fought, and won, cases against the biopiracy of our seeds and the indigenous knowledge of neem, basmati and wheat. As rural women sow the seeds of hope and resilience in times of multiple emergencies – health, hunger, climate and extinction – they grow gardens of hope for the health of the planet and of our communities.

*****

Ecology, economics and gender are all intimately connected to the construction of 'home' as a metaphor. The household was originally a metaphor for the economy, a word that has its roots in the Greek 'oikos', which referred to the family household and its daily operations and maintenance.

Before the emergence of the modern patriarchal paradigm of economies, it was assumed that national economies could be seen merely as extensions of the housekeeper's budget. With 'home' as the metaphor for both ecology and economics there was no hierarchical division between domestic production and commodity production for exchange and trade, or between nature's economy, the sustenance economy, and the market economy.

Modern economic paradigms, however, reconstituted the metaphor of home, separating it from the economy: home was redefined as the 'absence of economy'. A division of labour between the genders was also mapped homogenously, with women and men being projected onto the household and the economy, respectively. Sociologically, this led to what Maria Mies has called the 'housewifisation' of the domestic economy. As per patriarchal economic models, production in and for the home (and for need) is counted as 'non-production'. This transformation of value into non-value, labour into non-labour, knowledge into non-knowledge, is achieved by two very powerful constructs: the production boundary and the creation boundary.

The production boundary is a political construct which removes regenerative, renewable production cycles from the domain of production. National accounting systems which are used for calculating growth through Gross National Product (GNP) are based on the assumption that if producers consume what they produce, they do not, in fact, produce at all, because they fall outside the production boundary. Hence, all women who produce for their families, children and nature are treated as non-productive and economically inactive. The devaluation of women's work, and of work done in subsistence economies in the countries of the South, is the natural outcome of a production boundary constructed by capitalist patriarchy.

By restricting itself to the values of the market economy, the production boundary ignores economic value in two vital economies – the first is nature's economy, and the second is people's sustenance economy – that are necessary to ecological and human survival. In nature's economy, economic value is a measure of how the earth's life and human life are protected. Its currency is life-giving processes, not cash or the market price. The market economic value model has made women's work, and all domestic production, disappear in the blink of an eye. The exclusive focus on incomes and cash-flows (as measured in GNP) has meant that the web of life around women and the environment is excluded from central concern.

Among the hidden costs of a destructive development model are the new burdens created by an inherent ecological devastation, costs that are invariably heavier for women in both the North and the South. A rise in GNP or global trade figures does not necessarily result in a proportionate increase in either wealth or welfare. In fact, a very strong argument can be made to show that rising GNP or global trade figures are increasingly a measure of how real wealth, the wealth of nature and the sustaining wealth produced by women, is rapidly decreasing.

When trade in commodities is treated as the only economic activity, it destroys the potential of nature and women to enhance life and to produce goods and services for basic needs. More trade and more cash mean less life in nature through ecological destruction, and in society through denial of basic needs. As a result, women are devalued. First, because women, especially in rural and indigenous communities, work in cooperation with nature and her processes, which are often at odds with dominant market-driven development policies; second, because work that satisfies needs and ensures sustenance is devalued in general.

As the 'trade' metaphor replaced the metaphor of 'home', economic value itself underwent a change. Value which means 'worth', derived from *valere*, was redefined as exchange and trade; unless something is traded it has no economic value. The trade metaphor for the economy also rendered nature's economy as valueless.

This shift in economic value is central to the ecological crisis and is reflected in the changing meaning of the term 'resource'. 'Resource' originally implied life. Its root is the Latin verb *surgere*, which evoked the image of a spring that continually rises from the ground. Like a spring, a 'resource' rises again and again, even if it has repeatedly been used and consumed. The concept thus highlighted nature's power of self-regeneration and drew attention to her prodigious creativity. Moreover, it implied an ancient idea about the relationship between humans and nature – that the earth bestows gifts on humans who, in turn, are well advised to show diligence in order not to suffocate her generosity. In early modern times, 'resources' therefore suggested *reciprocity* along with *regeneration*.

With the advent of industrialism and colonialism, however, a conceptual break occurred. 'Natural resources' became those parts of nature which were required as inputs for industrial production and colonial trade. John Yeats, in *Natural History of Commerce*, offered the first definition of the new meaning in 1870: 'In speaking of the natural resources of any country, we refer to the ore in the mine, the stone unquarried, the timber unfelled.' In this view, nature has been clearly stripped of her creative power; she has been turned into a container for raw materials waiting to be transformed into 'input' for commodity production. Resources are now merely any material or conditions existing in nature which may be capable of 'economic exploration'. With the capacity for regeneration gone, the impulse for reciprocity has also lost ground: it is simply human inventiveness and industry that impart value to nature. Natural resources now need to be 'developed' – only when capital and technology have been introduced will nature find her destiny.

From then on, it became common sense that natural resources cannot develop themselves; it is only through the application of human knowledge and skill that anything can be made of them, and most of the necessary work requires skill of a very high order. Nature, whose real nature it is to rise again, was transformed by this worldview into inert and manipulable matter; its capacity to renew and grow was denied.

With the Uruguay Round of GATT and the establishment of the WTO the 'scope of the trade' metaphor increased further. Their two crucial impacts have been the removal of national boundaries in trade flows and the inclusion of subjects like agriculture and IPRs, which had hitherto been restricted to domestic policy, into the domain of international trade. Not only have the three economies of nature, people and the market been reduced to the market alone, the market economy itself has been reduced to the global market dominated by trade in dollars. This 'dollarisation' of the economic value of nature and of products necessary for human survival is an essential aspect of globalisation that has resulted in the reduction of local food consumption; of ground water resources; fisheries; agriculture; and livelihoods associated with traditional occupations in these sectors. For instance, intensive shrimp cultivation for export in south India has led to the disappearance of drinking water and of livelihood options. It is the women of the local fishing communities who suffered the most, and it is they who resisted the expansion of shrimp farming. As Govindamma of Kurru village in Nellore district of Andhra Pradesh puts it, 'We were displaced from the sea and went to agriculture for jobs. Now they are building prawn farms on agricultural land. Salt farms are also being converted to aqua. There, too, we will lose labour. Where will we earn our living?'

The TRIPS agreement brought the domain of ideas, knowledge and innovation into global trade for the first time. The creation boundary does to knowledge what the production boundary did to work. The separation of production from reproduction, the characterisation of the former as economic and the latter as biological, are underlying assumptions treated as 'natural', even though they are socially and politically constructed. This patriarchal shift in the creation boundary is misplaced for many reasons; the assumption that industrial activity is true creation because it takes place *ex nihilo* is ecologically false. No technological artefact or industrial commodity is created out of nothing; intellectual labour is expended at every stage of industrial production as 'raw material' or 'energy'. The biotech seed, which is treated as 'creation', to be protected by patents, could not exist without the farmer's seed. The patriarchal creation boundary allows ecological destruction to be misperceived as creation, and ecological regeneration and creation to be misperceived as non-creation.

Feminist analyses reveal that trade is yet another patriarchal project based on a gendered paradigm, with gender-differentialled impacts. Most women in South Asia are engaged in the vital sectors of agriculture, food processing, textiles and garments. These are also the major trade sectors in the region, and trade liberalisation leads to women losing both production options as well as consumption entitlements. An ecological and feminist agenda for trade must be based on recognising ecological limits and the social criteria that economic activity should be guided by if it has to respect the *environmental* principle of sustainability and the *ethical* principle of justice. This requires that the full ecological and social costs of economic activity and trade be made visible and be taken into account. Localisation, based on stronger democratic decision-making at local levels, building up to national and global levels, is an imperative for conservation as well as democracy. Localisation and decentralisation do not imply autarchy or isolation; rather, they are based on the principle of *subsidiarity*, and help to distinguish between different aspects of social and economic life, as well as in establishing the appropriate form of governance for each activity in order to ensure the protection of people and the environment.

Notes

[1] Late Professor Emeritus of Politics and Women's Studies at Mount Holyoke College, Massachusetts. She was also known as an activist for peace.

[2] Respected authority on biosafety, she was President/Director of the Edmonds Institute, USA.

[3] A biologist, author, researcher and activist who participated in the negotiations of the Rio Process on Sustainable Development, and was appointed expert for the major group 'Women' in the negotiations leading up to Rio+20.

[4] Health worker and founding member of International Public Health Coalition and People's Health Movement.

Against Appropriation

GATT & WTO

WE, IN INDIA, WHO HAVE BEEN UNDER COLONIAL RULE, ARE fully aware that neither globalisation nor free trade are new concepts. The first free trade treaty was signed by the East India Company in 1717 with the Mughal Emperor, Farukhsheer. We know that 'free' trade means unfair, asymmetric and unequal trade, which destroys local and national economies by privileging global corporations. The GATT is today's version of the Farukhsheer Firman. It was among the three institutions that were created at Bretton Woods in 1944 to continue colonisation in a post-colonial age. The other two were the World Bank and the International Monetary Fund (IMF). I became familiar with the World Bank during the 1980s, after doing an ecological assessment of eucalyptus monocultures in Karnataka, and later of its Tropical Forestry Plan. In 1988, when the huge mobilisations against the World Bank and IMF were organised in Berlin, a Russell Tribunal was organised to coincide with it, and I was invited to represent Nature. Teddy Goldsmith was my lawyer, defending Nature from the World Bank and IMF. I marched with citizens fighting for economic democracy and gave a speech at the historic Berlin Wall. I remember saying how walls like the Berlin Wall that divide people have no place in free and democratic societies. The Berlin Wall was pulled down the following year but new walls have come up since, including the invisible walls that the GATT/WTO have created.

Since 1987, I have closely followed trade issues, especially IPR agreements. GATT combined the creative aspects of design and copyright with patents on industrial products and processes and bundled them together as 'intellectual property' – property in products of the mind.

I was a regular visitor at our GATT embassy in Geneva when S.P. Shukla was our GATT ambassador. He firmly believed that patent laws and other intellectual property laws were sovereign legislation by national parliaments. But when GATT became WTO in 1994, the Trade Related Intellectual Property Rights agreement of the WTO became the instrument for the colonisation of nature. Our negotiators insisted that countries should be able to exclude plants, animals and seeds from patentability. B.K. Keayla of India's Patent Office set up a working group on patent laws; we facilitated an Interparty Parliamentary Group and worked together to include the amendment Article 3j in the Indian Patents Act, which states that plants, animals and seeds are not human inventions and hence not patentable. Bayer-Monsanto has repeatedly tried to challenge this clause but has failed. Prashant Bhushan has been our lawyer in the GMO case and on the Patents on Seed case. The National Working Group also introduced Article 3d, which states that trivial modifications in medicines are not 'inventions'. Bayer tried to challenge this clause in the case of a cancer drug; they were not successful. Anand Grover of the Lawyers Collective, who later became the UN Special Rapporteur on Health, has been leading the challenge against patents on medicine in partnership with public health experts like Dr Mira Shiva.

Chaturanan Mishra, India's agriculture minister in the late 1990s, invited me and S.P. Shukla to write the sui generis law on Plant Variety Protection. Since farmers are breeders and their right to exchange and save seeds is non-alienable, we drafted the law on Protection of Plant Variety and the Farmers' Rights Act.

The laws that Indian farmers are resisting today were first drafted in the early 1990s when the WTO imposed structural adjustments on India. When the Dunkel Draft Text was leaked, I travelled across India to talk to farmers about globalisation and the corporate control of agriculture. When I met Mahendra Singh Tikait, among the tallest national farmer leaders, and explained to him what was happening, he heard me out, smoking his hookah, then said, 'I understand; the East India Company is coming back. I am with you to fight them.' He was true to his word. We put together reports on the corporate hijack of biodiversity, of agriculture, of food, and of water to create awareness and build movements. Our reports on Monsanto and Cargill started major farmers' movements, including the farmers' actions against Cargill in 1998, and the Monsanto Quit India movement, launched on Quit India Day (August 9) the same year, when Monsanto illegally started its field trials of GMO Bt cotton.

After retirement, S.P. Shukla led the Peoples' Campaign against WTO, of which I was a part. Activists worked closely with leaders committed to people's rights as enshrined in the Constitution. We made a difference nationally and internationally. Globally, movements that were fighting GATT and globalisation came together as the International Forum on Globalisation (IFG). This initiative was started by Doug Tompkins and Jerry Mander. Among the founding board members, besides myself, were Maude Barlow and Tony Clarke, who wrote *Blue Gold* and received the Right Livelihood Award for their work on the privatisation of water; Martin Khor of the TWN; Walden Bello[1] of the Philippines; Teddy Goldsmith; and Sara Larraín[2] of Chile. Others in the IFG group included Tim Lang and Colin Hines; David Korten, who wrote *When Corporations Rule the World*; and Ralph Nader and Lori Wallach of Public Citizen. We gathered in San Francisco twice every year to learn from each other, share our experiences and create a coherent movement across the world to protect the planet, people's rights and livelihoods. The Forum allowed us to tell the real story of the impact of globalisation in our countries. We wrote *Alternatives to Economic Globalization*; we organised teach-ins and protests during the 1999 WTO Seattle meeting with the slogan 'Our World Is Not for Sale'. Environmentalists marched along with trade unionists; farmers of the North and the South spoke in one voice. People caused the WTO Seattle Ministerial to stop, demonstrating that when people are united, they have the power.

The WTO is in intensive care. The 2001 Doha Ministerial was held under high security; but despite the severe restrictions, movements were able to bring the issues of farmers' suicides and the monopoly control of corporations over seeds and medicines to the fore. In July 2001, the violence of globalisation took the life of young Carlo Giuliani, who was shot dead by the police during an anti-globalisation protest at the G8 Summit in Genoa. The WTO failed in Cancun in 2003 and barely survived in Hong Kong in 2005. The issue of vaccine patents and vaccine apartheid, too, has been intensely debated, but the Big Food and Big Pharma – and they are the same – that wrote the TRIPS agreement to have monopoly over seeds and medicines continue to rake in profits at the cost of people and the planet.

Klaus Schwab[3] of the World Economic Forum (WEF), a club of the richest corporations, invited me to Davos in 2000 to address his corporate gathering on globalisation. I debated Dupont and the World Bank on that occasion. In 2000, the WEF also invited me to their conference in Melbourne, held from September 11-13. On September 11, tens of thousands of people gathered to speak up against the corporate rule of the World Bank, the WTO and the WEF. I marched with them. On September 12, I took their permission to go to the WEF to debate Bill Gates, who was forced to arrive in a helicopter because the protestors had blocked the roads. I went up on stage and read out the letter from the protestors, who the corporate media had labelled as a bunch of 'trouble-causing hecklers':

> We are students, mothers, fathers, workers, the unemployed, [and] environmentalists, from religious traditions, indigenous people, amongst others. We are blockading the Crown Casino today to use one of the tools that we, as community groups and individuals, have – civil disobedience.
>
> We are blockading because people across the world are suffering under corporate-defined globalisation. Many of these people are denied the opportunity of civil disobedience because of military regimes, some of which receive corporate support. Unlike the corporations represented inside the World Economic Forum, we do not have multi-million dollar advertising budgets or public relations

consultants. What we have is a steadfast belief that we have both the right and the responsibility to take action in the face of corporate disregard for human rights, environmental protection, public health and labour regulation.

This is something we – here in Australia and around the world – have stated for many years. While the WEF claims that it is now addressing key issues such as human rights and environmental sustainability, this is not reflected in the behaviour of individual WEF members and we see no evidence of equality in the distribution of global wealth.... We are part of a worldwide movement demanding justice before profits. We are not going away.

In 2001, I was invited back by the WEF to Davos to advise them on issues of globalisation. There, too, protestors had gathered outside the WEF meetings and they asked me to come and address them. As I walked across the snow to meet them, a policeman tried to strike me with a baton. I managed to join the protestors, but on returning to the conference hall I held a press conference on the duplicity and double standards of the WEF: inviting us to give speeches as experts on globalisation inside the Forum whilst declaring us criminals when we organised as movements outside. I demanded an apology, failing which I would never return to the WEF. Schwab did not apologise – and I have never returned to the WEF. But I still follow how, through the WEF, the corporations and billionaires are hijacking democracy. The WEF and Bill Gates hijacked the food summit for the corporate agenda of farming without farmers; food without farms; patented, synthetic lab food; digitalisation of every sphere of our lives for surveillance, extraction and profits. Our work is even more significant now as the destructive power of billionaires and corporations increases. Our work is to create alternatives – living, local, circular economies in which all life on earth and all people can thrive.

*****

While the risks of biopollution and genetic pollution created by GMOs are well-known and well-established empirically, and even though the CBD and the Biosafety Protocol created a legal framework for addressing issues related to the impact of GMOs on the environment and public health, the US has time and again tried to subvert the discussions on biosafety by moving them from where they belong, viz., the CBD, to the WTO, which neither has the mandate nor the capacity to create environmental regulations.

The objective for moving biosafety issues from a multilateral environment agreement, the CBD, to a free trade agreement, the WTO, became evident in Para 29 (vi), under the Section on Agriculture of the Ministerial Text of October 19, 1999, for the WTO meeting in Seattle. It called for 'disciplines to ensure that trade in products of agricultural biotechnology is based on transparent, predictable and timely processes'. It was ironic that the country that had prevented transparency in the trade of GMOs, by refusing segregation and labelling of GM products, was referring to 'transparency'. It was clear that the US did not mean transparency in the context of the democratic rights of citizens; it wanted to ensure transparency for corporations, essentially implying their easy access to government decisions.

The challenge at Seattle, therefore, was to stop the further deregulation of GM trade, characterised by political and environmental unaccountability, and halt the trend that was transforming environmental problems, which need environmental solutions, into trade problems, with further trade liberalisation and deregulation offered as solutions. The challenge was to create a new political context in which people, not corporations, set the agenda.

The Citizens' Agenda at the WTO Seattle Ministerial Conference wanted to:

1. Prevent the acceptance of the US proposals on new free trade disciplines for GM trade.
2. Prevent the setting up of a 'working party' or 'examination group' under the WTO as proposed by Canada and Japan. Such a working party should instead be set up under the CBD, which has the mandate to deal with the impact of GMOs on the environment.
3. Call for a Five Year Freeze on GM trade while further ecological research and health research are carried out on the environment and health impacts of GMOs; the biosafety protocol is finalised; and strong biosafety regulations are implemented in all countries.
4. Prevent the World Bank from using development funds for promoting GM production and trade in the Third World.
5. Stop the dumping of GM foods on the poor by preventing food aid funds from being used in order to create markets for GM foods.

The battle for Seattle

I had been fighting the GATT/WTO since the late 1980s because the corporations that drove the agenda at GATT talked about patenting seeds and life.

The Poison Cartel, which controls our seeds and our food, is the product of the WTO. Bill Gates is a child of the WTO. He had its rules formulated so that he wouldn't have to pay taxes during transborder transfer, which is why software was outsourced to India. Jeff Bezos of Amazon ships goods around the world without paying taxes anywhere. These trillionaires are the children of WTO rules.

The Third WTO Ministerial Conference in Seattle was a historic watershed for the anti-globalisation, anti-GMO movements worldwide. I was in Seattle as part of the IFG. We conducted several teach-ins in the run-up to the Ministerial meeting, and while we had anticipated the participation of around 30,000 people, instead thousands more turned up. Young people on the streets came up to me and told me how they were there because of biopiracy. They were there to stop the privatisation of water; they were there to defend our public goods; they were there to defend our commons. Everyone was speaking in one voice.

The conference was a failure, and I was not surprised at the manner in which it was shut down. It only demonstrated that globalisation is not an inevitable phenomenon which must be accepted at all costs; rather, it is a political project which can be responded to politically.

Around 50,000 citizens from all walks of life and all parts of the world were responding politically when they protested peacefully on the streets of Seattle for four days to ensure that there would be no new round of trade negotiations for accelerating and expanding the process of globalisation.

Trade ministers from Asia, Africa, Latin America and the Caribbean were responding politically when they refused to join hands to provide support to a 'contrived' consensus because they had been excluded from the negotiations being undertaken in the 'green room', behind closed doors. As long as conditions of transparency, openness and participation were not ensured, developing countries would not be party to a consensus. This was a new context and made the bulldozing of decisions difficult for future trade negotiations.

The rebellion on the streets as well as the rebellion within the WTO started a new democracy movement – with citizens from around the world and the governments of the South refusing to be bullied and excluded from making decisions in which they have a rightful stake.

Seattle was chosen by the US to host the Third Ministerial Conference because it is the home of Boeing and Microsoft, and symbolises the corporate power which the WTO rules are designed to protect and expand. Yet, the corporations stayed in the background while the proponents of free trade and the WTO went out of their way to say that the WTO was a 'member-driven' institution controlled by governments who made democratic decisions. The refusal of the governments of the South to rubber-stamp decisions from which they had been excluded brought into the open, and confirmed, the non-transparent and anti-democratic processes by which WTO rules are imposed on the countries of the global South.

The WTO had earned epithets such as World Tyranny Organisation for itself because it enforced tyrannical, anti-people, anti-nature decisions to enable corporations to steal the world's harvests through secretive, undemocratic structures and processes. It institutionalised forced trade, not free trade, and beyond a point, coercion and the rule of force cannot continue.

The WTO tyranny was apparent in Seattle both on the streets and inside the Washington State Convention Centre where the negotiations were taking place. Nonviolent protestors, including young people and old women, labour and environmental activists, and even local residents, were brutally beaten up, sprayed with tear gas and arrested in their hundreds. The intolerance of democratic dissent, which is the hallmark of dictatorship, was unleashed in full force in Seattle. While trees and stores were lit up for Christmas, the streets were barricaded and blocked by the police, turning the city into a war zone.

The media referred to the protestors as 'power mongers' and 'special interest' groups, while globalisers, such as Scott Miller of the US Alliance for Trade Expansion, said that the protestors were acting out of fear and ignorance. But the thousands of youth, farmers, workers and environmentalists who marched in peace and solidarity were not acting out of ignorance and fear; they were outraged because they understood how undemocratic the WTO is, how destructive its social and ecological impacts are, and how its rules are driven by the objective of establishing corporate control over every dimension of our lives – our food, our health, our environment, our work and our future.

When labour joins hands with environmentalists, when farmers from the North and the South make a common commitment to say NO to genetically engineered crops, they are not acting in their special interest. They are defending the common interests and the common rights of *all* people, everywhere. The divide and rule policy, which has attempted to pit consumers against farmers, the North against the South, labour against environmentalists, failed in Seattle. In their diversity, citizens were united across sectors and regions.

While the broadbased citizens campaigns stopped a new Millennium Round of the WTO from being held in Seattle, they did launch their own Millennium Round of the democratisation of the global economy.

This real Millennium Round marked the beginning of a new democratic debate about the future of the earth and its people. It was declared that the centralised, undemocratic rules and structures of the WTO, based on monopolies and monocultures, need to give way to an earth democracy supported by decentralisation and diversity. That the rights of all species and of all people must come before the right of corporations to make limitless profit through unbridled destruction. That free trade is not leading to freedom, it is leading to slavery – diverse life forms are being enslaved through patents on life, farmers are being enslaved into high-tech slavery, and countries are being pushed into debt, dependence and the destruction of their domestic economies.

We want a new millennium based on economic democracy, not economic totalitarianism. The future is possible for humans and other species only if the principles of competition, organised greed, commodification of all life, monocultures, monopolies and centralised global corporate control of our daily lives, enshrined in the WTO, are replaced by the principles of protection of people and nature, the obligation of giving and sharing diversity, and the decentralisation and self-organisation enshrined in our diverse cultures and national constitutions.

A new threshold was crossed in Seattle, towards the creation of a global citizen-based and citizen-driven democratic order. With the slogan 'No New Round, Turnaround', the protestors were successful in blocking a new round.

The Citizens' Seattle Round synthesised the common concerns of people from across the world to ensure that the way we produce, distribute, process and consume food is sustainable and equitable. In the South as well as in the industrialised North, common principles have emerged from people's practices to ensure a safe and healthy food supply. These principles enable us to shift to nature-centred and people-centred food systems:

- Diversity rather than monocultures to ensure higher output per acre.
- Decentralisation and localisation in place of centralisation and globalisation.
- Ecological processes instead of industrial processes of farming.
- Food rights and food security rather than free trade as the basis of distribution.
- Democratic control rather than corporate control of the food system.

- Patent-free and genetic engineering-free farming to ensure the respect and protection of all species and the integrity of ecosystems and cultures. This involves excluding life forms from TRIPS and biosafety from WTO rules of free trade.
- Cultural diversity in place of the global monoculture of fast foods and industrial food chains.
- Small farms and small farmers in place of corporate farms and absentee landowners. This involves protection of existing small farms and land reforms to redistribute land.
- Fair trade, not free trade, to ensure farmers and producers get a fair return. Trade as a means rather than an end, with global trade subservient to values of ecological sustainability, health and social justice.

Against all odds, millions of people across the world have been putting these principles into practice. The post-Seattle challenge has been to change global trade rules and national food and agricultural policies so that these practices can be nurtured and disseminated, and that ecological agriculture, which protects small farms and peasant livelihoods and produces safe food, is not marginalised and criminalised. Seattle reiterated the need to reclaim the stolen harvest and celebrate the growing and giving of good food as the highest gift and the most revolutionary act.

What we need to learn from Seattle is that when people are determined, the Turtles and the Teamsters can walk together to defend the rights of the earth and our collective rights. Our world is on the verge of destruction and climate catastrophe, because those who make money out of destroying the planet want to continue doing so. That's the moment we are in today. We need to unite to fight for the planet and fight for every last person, including the last displaced person, who is today's refugee.

*****

For more than three decades Indian farmers have resisted the World Bank–imposed Structural Adjustment of 1991 and the GATT/WTO rules of corporate free trade. This included conditionalities that would dismantle the Essential Commodities Act, which prevents hoarding foodgrains and speculation, and the Agricultural Produce Marketing Committee (APMC) Act that regulates the diverse markets in India from local village markets to the regional wholesale markets. I call the Essential Commodities Act the Prevention of Famine Act and the APMC an act that guarantees fair trade.

During the height of the COVID-19 pandemic and lockdowns, the Indian government drew up a slew of agricultural 'reforms' to further the agenda of the World Bank. Three major farms laws were introduced in parliament in September 2020:

1. The Essential Commodities Act for regulating stock-piling food and preventing speculation was amended to *exclude* food through the Essential Commodities (Amendment) Act, 2020.
2. State laws for regulating markets and traders were done away with through the Farming Produce Trade and Commerce (Promotion and Facilitation) Act, 2020.

3. A contract farming law, titled The Farmers (Empowerment and Protection) Agreement on Price Assurance and Farm Services Act, 2020, opened the doors for global agribusiness giants, e-commerce corporations, and food processing companies to trap farmers into new corporate slavery. Transnational corporations like Bayer-Monsanto, Cargill, PepsiCo, Amazon and Walmart are partnering with Indian corporates to capture the domestic market, emerging as the new zamindars who control not just our food and land use, but also our water and seeds.

However, parliamentary democracy and people's democratic movements prevented the realisation of this blatant corporate agenda. For over twelve months, from October 2020 to December 2021, thousands of protesting farmers camped at the Delhi border demanding a repeal of the three laws, a death knell for our small farmers and a threat to our food sovereignty. More than 600 farmers were martyred during the year-long struggle, which I call the Third Movement of India's Freedom, showing the path to food freedom for all of humanity.

The courage, solidarity and unity of our farming communities, as well as the organisational creativity and tirelessness of their nonviolent resistance, offer lessons for movements around the world. This is not just a movement for farmers' freedom; it is a movement for defending human freedom.

The strength of the movement forced the government to announce a repeal of the three laws on November 19, 2021. On November 20, I was invited to give the keynote address at a national convention on 'Farmers' Struggle and Earth Democracy', organised by the Punjab Women Collective, comprising eleven organisations, to pay tribute to the women martyrs of the movement. Others present at the event included Medha Patkar of the Narmada Bachao Andolan and writer–activist Dr Navsharan Kaur among the key speakers, as well as Devi Kumari,[4] Jasbir Kaur Nat,[5] Paramjit Longowal,[6] and Bhagat Singh's niece, Gurjeet Kaur.

The convention was organised one day after the announcement, and it became a platform for celebration as well as for discussing a plan for food freedom for the next phase of globalisation: digital agriculture; new GMOs; fake food; fake accounting for a new carbon colonisation as a fake solution to climate change; and the financialising and monetising of nature by corporations.

Prime Minister Modi chose the auspicious day of Guru Purab, the birth anniversary of Guru Nanak, the founder of the Sikh religion, to make his announcement. Guru Nanak saw humanity as part of nature, and the protection of nature as the highest duty of human beings. The Sikh langar (community kitchen) has fed people for centuries, and continues to do so even today, especially during an emergency, be it the war in Syria or the COVID-19 health crisis. As I said at the Convention,

> The Green Revolution 'chemicalised' Punjab. The State must practice agriculture as Guru Nanak would have liked. Organic is not a luxury, it's our duty. Since half of India cannot pay for food, it should be supported by the Public Distribution System (PDS); it should be the national langar.

The attack on our PDS by the World Bank was intense. The globalisation of agriculture and agricultural revolution don't go together.

The farmers' struggle will continue till all their demands are met, including a law to guarantee a fair price, a minimum support price, an equivalent in terms of minimum wages for workers. The movement has shown that people united can never be defeated. Muslims and Hindus sat together to defend our food freedom and food sovereignty; women and men fought side-by-side; landless workers and landowners came together in a new show of unity as cultivators and caretakers of the earth, whom Nanak called the 'great mother of all'; singers and artists joined in. 'No Farmers, No Food' was the slogan that this movement raised, an important call to resist the privatisation of food.

Notes

[1] Filipino academic, environmentalist and social worker who served as a member of the House of Representatives in the Philippines.

[2] Chilean politician and environmentalist who ran for president in the 1999 presidential election.

[3] German engineer and economist, best known as the founder and executive chairman of the World Economic Forum (WEF).

[4] Member of Punjab Khet Mazdoor Sabha.

[5] Member, National Executive, All India Progressive Women's Association (AIPWA) and State Committee Member, Punjab Kisan Union.

[6] The general secretary of Zamin Prapti Sangharsh Committee (ZPSC), a Dalit rights group in Punjab.

The Great
Water
Capture

THE TEHRI DAM, BUILT ON THE BHAGIRATHI RIVER IN Uttarakhand, was planned in the 1960s to 'create a reservoir for irrigation, water supply and hydroelectricity'. However, this mega project has met with stiff opposition ever since its initiation, with protests against the displacement of Tehri's inhabitants as well as the environmental consequences of locating such a large dam in the fragile ecosystem of the Himalayan foothills.

In 1969, the Member of Parliament from Tehri, Rajmata Kamlendu Mati Shah, raised strong objections to it. In 1972, 35 villages in Tehri district passed a resolution asking the government not to build it. In 1978, eminent freedom fighter, lawyer and former MLA of Tehri, Veerendra Saklani, set up the Anti-Tehri Dam Struggle Committee with Sunderlal Bahuguna as a member and spokesman. The committee organised a peaceful protest, and along with Veerendra Saklani, 150-200 satyagrahis were arrested and sent to various jails, including Bareilly jail.

During my frequent visits to Silyara and Tehri, I would stop to meet Veerendra Saklani and his wife. Every evening Mrs Saklani would light a lamp to Ganga Ma in a small temple on the banks of the Bhagirathi. Her love of, and faith in, the Ganges was deep and steadfast.

In 1980, the then Prime Minister, Indira Gandhi, wrote to the Department of Science and Technology about the need to reconsider the Tehri Dam Project. In 1986, the Chairman of the Working Group appointed by the department recommended that work be stopped on the dam. The Ministry of Environment and Forests also wrote to Rajiv Gandhi, then Prime Minster, to abandon the project, but construction continued.

The protests intensified. Slogans reverberated across Tehri.

Implement the recommendations of the scientific committee. Stop the Dam.

No one will move, no one will be drowned/We will stop the dam.

The satyagrahis also painted slogans on the mountainside:

Tehri Dam is a symbol of total destruction.
It is against the national interest and against nature.
Protect the environment.
Stop the dam.

On November 24, 1989, Sunderlalji took a pledge to not return to Silyara Ashram until the dam was stopped. On Christmas day, in 1989, he embarked on a fast to stop the construction and perched on a bulldozer's wheel. His fast lasted 16 days. He built a Satyagraha Hut near a temple 100 metres away and said that if the dam was not halted he would offer himself to the Ganga in a jal samadhi (death by water). I visited Sunderlalji frequently in his Satyagraha Hut, which had become his and Bimla Bahuguna's new Ashram. In 1991, a disastrous earthquake occurred, 6.6 on the Richter scale, with its epicentre at the site of the Maneri Dam on the Bhagirathi. The Silyara Ashram was destroyed, along with 28,000 homes, and 770 people died.

On December 14, 1991, 5,000 satyagrahis gathered at the dam site where they set up tents, and Bimlaji and the women sat on a dharna as explosives were blasted to create the tunnels for the dam. Work on the dam stopped for 75 days. At midnight of January 27-28, 1992, armed police uprooted the satyagrahis' camp, and 40 satyagrahis were bundled into police vans. Sunderlalji and Bimlaji were dragged out of the camp, arrested, and taken to Roorkee jail. Both of them undertook a hunger strike in protest; Bimlaji broke hers after eleven days, but Sunderlalji continued with his till March 6, 1992, when the court ordered his release.

For the next few months the situation remained fluid as the satyagrahis continued their protest; the government commissioned more reports regarding the dam's safety, but despite warnings by scientists and seismologists, work on the dam's construction resumed.

On June 9, 1995, at 3:00 a.m., Sunderlalji was pulled out of his hut, driven to Jolly Grant Airport in Dehradun, flown by helicopter to Delhi, and deposited in the Emergency Ward at the All India Institutes of Medical Sciences (AIIMS). It was an arrest of sorts.

The People's Union for Civil Liberties filed a case in the Allahabad High Court challenging his detention in AIIMS. The government's lawyer replied that his long fast had placed Sunderlalji's life in danger. But the Court ordered his release and he returned to the satyagraha site to continue his fast.

Despite the people's movement attracting national and international attention during the 1980s and the 1990s, the construction of the dam continued – albeit with several interruptions – and eventually the rising waters of the Tehri Dam submerged not only the old Tehri town but also many villages and farmlands upstream on the Bhagirathi and Bhilangana rivers. By 2004, the first phase of the construction was completed, although there were protracted legal battles over the resettlement rights of more than 100,000 people to the town of New Tehri. Since 2005, the filling of the reservoir has led to the reduced flow of Bhagirathi water, and while this reduction has been central to local protests against the dam, officials claim that when the reservoir is filled to its maximum capacity the flow of the river will become normal. In spite of concerns and protestation, the operation of the Tehri Dam continues – its 1,000 mW variable-speed pumped-storage scheme is currently under construction.

*****

I have grown up with free-flowing, unpolluted streams and rivers, at a time when we could stop and drink water from any source. Our rivers were not dirty drains carrying pollution and waste, including urban waste. Pesticides had not seeped through, so ground water was clean and potable in the poorest village, and even in the desert, water was available at 10 ft. The Persian wheel and indigenous water-lifting technologies were scattered across the rural landscape.

Then the World Bank brought in 'development' and created a water crisis! It forced 'liberalisation' of the Indian economy with its 1991 loan, and the WTO 'free trade rules' made structural adjustment a permanent process. The privatisation and commodification of water lay at the heart of the World Bank programmes and the WTO trade rules.

In 1991, when the World Bank was imposing its structural adjustment programme on India, I got in touch with Dr Banwarilal Sharma, a maths professor at the University of Allahabad, and joined hands with him to start the Azadi Bachao Andolan (the Movement to Defend Freedom). George Fernandes and Rabi Ray, erstwhile member Speaker of our Parliament, were part of the founding of this movement in Sevagram. I also worked with Dr Sharma on the Ganga Satyagraha, to stop the privatisation of the Ganga by the Suez corporation with the help of the World Bank.

Rivers and streams across the country are being drained of water by private companies, and even the Ganges has not been safe from this kind of privatisation. Suez, the world's largest water corporation, set up a plant in Delhi at Sonia Vihar to sell 635 million litres of Ganges water to the affluent of Delhi. The Gangetic plain is one of the most fertile regions in the world. At the beginning of the ploughing season in Bihar, for example, farmers place Ganges water in a pot and set it aside in a special place in the field before they plant their seeds, to ensure a good harvest. ONDEO-Dégremont, a subsidiary of Suez Lyonnaise des Eaux (Water Division), was awarded a Rs 2 billion contract (approx. $50 million) for the design, construction and operation (for 10 years) of a 635 million litres per day drinking water production plant at Sonia Vihar; it was inaugurated on June 21, 2002.

Suez is a familiar name in water privatisation, along with Vivendi, Bechtel and Nestlé. Having seen their announcement, I began investigating, and got in touch with the workers' union at Delhi Jal Board; they provided me with more information on the World Bank project and the PriceWaterhouse consultancy report on it. I did some more research on how the water would be brought from the Ganges at Tehri Dam to the Sonia Vihar plant. We built a

Water Democracy Movement, connecting the people who had been displaced by the Tehri Dam, the farmers along the Ganga Canal, the people from slum colonies who would lose their supply of public water, and Resident Welfare Associations. We organised a referendum in Delhi against water privatisation by collecting signatures on a kilometre-long cloth to hand over to the government. We organised a Ganga Yatra from Tehri to Delhi, a satyagraha at Sonia Vihar, conferences to demand the cancellation of the World Bank project, and the rejuvenation of the Yamuna to sustain Delhi's water needs. We invited then Chief Minister of Delhi, Sheila Dikshit, to our protest.

From March 12-14, 2003, diverse movements related to water conservation and water rights undertook a Ganga Yatra from Tehri to Delhi. In the group were Sunderlal Bahuguna, who had been sitting at the Tehri Dam site as a witness to the destruction of the Ganga; Oscar Olivera of Bolivia who, with his fellow citizens, drove out Bechtel and regained social control over Cochabamba's water; and Ramon Magsaysay Award winner Rajendra Singh, of the organisation Tarun Bharat Sangh.

During our Yatra we saw that villages in the catchment of the Tehri Dam were systematically being denied water. All pumping schemes were cancelled because they would 'reduce' the water in the Dam. One hundred women committed suicide due to water scarcity during those years. Women from 250 villages told us that their sacred mother, the River Ganga, had been reduced to receiving them in death and could no longer give them life; but they refused to give up their struggle, saying they would commit collective suicide if forced to move. With them were Bimla and Sunderlal Bahuguna.

On this Yatra we also saw how farming in the richest soils of the world, the Doab between the rivers Ganges and Yamuna, was being threatened. The disastrous impact of diverting Ganga water to Delhi on the farmers of western UP was obvious, because this area is totally dependent on the Canal for irrigation. Even before their water supply was usurped, the farmers had begun to feel the impact of corporate greed: the government lined the Upper Ganga Canal to prevent seepage into the neighbouring fields (an important source of moisture for farming and for recharging ground water); and farmers were prevented from digging wells, even though they were reeling from severe drought.

On August 9, 2002, more than 5,000 farmers of Muradnagar and the adjoining areas of western UP had gathered in a rally in the village of Bhanera to protest the laying of a giant 3.25 metre-diametre pipeline for the Ganga project. This rally was the culmination of a 300-kilometre-long mobilisation drive along the Ganga by the farmers of Garhwal and the inhabitants of the devastated city of Tehri, to liberate the river from privatisation. The rally was launched from Haridwar where hundreds of farmers, together with priests and citizens, announced that the Ganga is not for sale. Thousands of farmers and others in villages along the route joined the rally.

At a meeting organised by Navdanya in Chaprauli on July 21, 2003, farmers said they would not allow the Canal to be lined to supply water to Delhi. Instead, the government should link the Upper Ganga Canal to the Yamuna Canal passing through this area to tackle the severe drought.

Throughout 2003, the Year of Fresh Water, a Jal Yatra, or water journey, was undertaken across India to recover water and the rivers as common property resources, as well as to conserve water by shifting to water-prudent crops and farming methods, and to resist privatisation.

Before the World Social Forum took place in 2004 in Mumbai, we organised a Conference on Water Privatisation in Delhi to support Delhi's Water Democracy Movement. Danielle Mitterrand, widow of François Mitterrand, who founded France Libertés, joined us, as did other activists from around the world who were fighting all World Bank water privatisation projects.

When the then President of the World Bank visited Delhi, he asked to meet me; I said I would, but would also bring others who were part of the movement with me. Women with empty water-pots gathered at the World Bank building in Lodhi Estate, near the Lodhi Gardens. I went inside with Dr Ramaswamy Iyer, who had been water secretary in the government and was committed to defending water as a commons and a public good. I had carried the movement's demand for cancellation of the World Bank water privatisation project with me. I also carried a small copper pot with Ganga water in it. While leaving, I told the President, 'Keep this by your bedside to remind you, when you go to sleep every night and when you wake up, that the Ganga is our sacred mother. She is not a commodity. Mother Ganga is not for sale.' The World Bank project that proposed restructuring the water system as well as water tariffs had to be cancelled.

Some people argue that one plant in Delhi does not imply 'privatising' the Ganga. However, the 635 million litres which Suez diverted from the Ganga to the Yamuna basin offer ecological services in the Ganga basin. To divert the water, the traditional rights and access to drinking water and irrigation by local communities in Uttaranchal and the Doab region of UP were denied. In effect, the privatisation of distribution translates into ownership and control of the resource, and the commodification of the basis of life.

Among the mega projects we challenged was the one that proposed linking all of India's rivers through super dams and super canals, at a cost of $200 billion − two hundred times what India spends on education; three times what the government collects in taxes; 25 per cent of our GDP; and US $72 billion more than India's total external debt. This mega project offered new opportunities for privatisation, but entailed cruelly heavy costs for our rivers and people. It meant the displacement of more than five million people, with ecosystems and communities being deprived of water, imprisoned by giant dams and canals of cement. Free-flowing rivers would be transformed into captive water; people with free access to their bathing ghats, wells, tanks and ponds would instead be bonded to giant water companies and water bureaucracies. This was a nightmare of slavery, a recipe for the extinction of species and cultures.

*****

In India, human settlements are called *abadi* − which means that they are based on water. The word for water is *ab* in

Persian and *ap* in Sanskrit. *Abroo* means human dignity – it is the spirit of water. The right to water is a natural right, a birthright, and is common to all beings. Common rights go hand in hand with common responsibilities; the culture of conservation and the relationship between common rights and responsibilities has supported human life, and all life on earth, for millennia.

Maldevelopment, which has increased commerce but decreased life's renewable potential by creating huge social and environmental externalities, has left us with polluted rivers, depleted ground water, desertified soils, and thirsty people. The solution offered by the World Bank/IMF and the WTO to combat the water scarcity created by non-sustainable development and water injustice is privatisation, or private-public partnerships. Unfortunately, privatisation leads to accelerating the non-sustainable use of water and deepening the hydrological divide. It leads to corporations controlling water supplies, and the wealthy buying more than their fair share. As a result, those without purchasing power are denied their water rights, and hence their right to life. Privatisation of water polarises society; it is the ultimate human rights violation, the ultimate human wrong.

Water privatisation has been justified by the logic of 'full cost accounting', i.e., that private companies should be able to recover the full cost of their investment when water giants get access to water markets through water privatisation. However, as we will see from the following examples, corporations actually obtain water for free, without paying the full social and environmental costs to those rural communities from whom the water has been taken. In India, even the funds for privatisation studies, which came from a World Bank loan of $2.5 million, did not go directly to the people; of the total amount, $1.9 million was paid as consultancy fees to PriceWaterhouse – the remaining $600,000 was spent on providing facilities to private consultants in India.

One outcome of the World Summit on Sustainable Development (WSSD), held in 2002, was the public-private partnership project entitled, Washing Hands. This project – launched by the World Bank, the London School of Hygiene and Tropical Medicine, USAID, UNICEF, WHO, as well as detergent companies like Unilever, Procter & Gamble and Colgate-Palmolive – talked of 'saving lives' by reducing diarrhoeal diseases by half. How? By doubling hand-washing and selling soap! Kerala was selected for implementing the Washing Hands project in India – even though the state has the highest hygiene standards, the lowest diarrhoeal deaths, the highest awareness on prevention of diarrhoeal diseases, the lowest childhood mortality, and the highest female literacy rates in the country. Most importantly, Kerala has the highest access to safe water. The World Bank project was an insult to Kerala's knowledge and practice regarding health and hygiene. In fact, it is Kerala from where cleanliness and hygiene should be exported to the rest of the world. The people of Kerala did not need this World Bank loan to be taught proper hygiene; clearly the project was not about 'saving lives', merely about 'selling soap'. It was about using the propaganda machine to discredit the state's health and hygiene standards.

Kerala has the richest indigenous resources for non-chemical, non-polluting natural hygiene products, from herbal soaps to natural soap-making at the small-scale level.

The Washing Hands project was simply an attempt to dismiss indigenous knowledge, indigenous biodiversity and indigenous economies, seriously endangering local cottage industries, as well as introducing chemical-based toxic detergents as pollutants into the system. What was most worrisome was that the project was being launched to legitimise water privatisation through private-public partnerships, aimed at undermining people's water rights and the state's duty to protect them. As with Coca-Cola's presence in Kerala, private-public partnerships have often been a recipe for over-exploitation of scarce fresh water resources.

Water wars in Plachimada

Coca-Cola was thrown out of India in 1977 by the then industries minister, George Fernandes, because of financial fraud. When it returned in the 1980s, it started setting up bottling plants everywhere. I had never liked the taste of the brown liquid, but it was only in 2002 that I realised the high social and ecological costs that this so-called 'soft drink' and the bottled water industry entailed.

That year, out of the blue, I received a call inviting me to celebrate Earth Day (June 22) with women in Plachimada who were resisting a Coca-Cola plant in their village. I did not know where Plachimada was, but I wanted to know why village communities were fighting Coca-Cola. So I took a flight to Coimbatore and drove four hours by taxi to Plachimada, which is outside Palghat in Kerala. On August 4, 2002, 300 adivasis (tribals) of the Coca-Cola Virudha Samara Samithy (Anti Cola-Cola Struggle Committee) were arrested at a mass rally at Plachimada for protesting Coca-Cola's appropriation of the village's common water resources for its water-bottling plant.

I spoke to the gathering about celebrating and saving Mother Earth, and renewing my commitment to defending water as a commons. Mylamma, a tribal elder, who was leading the movement, honoured me with a handmade bow and arrow. When I asked her what message she wanted to send to the people of Delhi and the world, she said, 'Tell them, when they drink Coca-Cola, they drink the blood of my people.' I have not been able to touch a Coca-Cola product since that day. No matter how thirsty I am, if someone brings me a bottle of Kinley, the bottled water brand from the same company, I gently refuse.

Later, we invited Mylamma and honoured her at a women and water conference in 2004 that we organised at the Earth University at Navdanya. Mylamma is no more, but the Coca-Cola plant in Plachimada was shut down in her lifetime.

While in Plachimada, I called Veerendra Kumar, then a Member of Parliament and President of the Janata Party. I know Veerendra Kumar because of our common work against the WTO and globalisation in Parliament and in Kerala. He owned the Malayalam daily newspaper *Mathrubhumi*, which had been started by his family as their contribution to India's freedom movement. I talked to him about the women's struggle and said that his paper should cover their protest and also raise it politically. Coca-Cola withdrew Rs 6 crores in advertising from *Mathrubhumi* after it published the Plachimada story, but Veerendra Kumar refused to back down. We organised rallies and conferences, and initiated court cases; *Mathrubhumi*'s courage and strong stand forced other papers to also cover the women's resistance to Coca-Cola. At one of the conferences, I invited fellow water warriors from the global movement against privatisation – Maude Barlow;[1] José Bové, the French cheese farmer who bulldozed the McDonald's outlet in his town in southern France (after the US blocked cheese and wine imports from Europe as part of trade wars); and parliamentarians from the European Parliament. The Satyagraha Hut, where women sat in protest every day, was opposite the Coca-Cola factory gate. We joined the women in the hut and then marched collectively to the gate. The management came to invite us in to see the plant, but without local representatives. We refused. We said we were guests of the local community and would go inside the premises only if they could accompany us. Coca-Cola was forced to allow local representatives to lead our delegation. They showed us a propaganda film; the movement representatives presented facts. We returned to the gate and continued our protest, and the women continued their satyagraha till the closure of the plant in 2004.

Coca-Cola had been drawing 15 lakh litres of water per day, which dried up the aquifers over a two-year period and polluted the water. Water scarcity hit the local adivasi and Dalit community the hardest. Adivasis asserted their right to water and demanded that Coca-Cola restore the environment, pay compensation, close down the factory, and leave the country. Coca-Cola's plunder was not restricted to Kerala; in Ghaziabad (UP), the company drew about 200 cusecs of water every day through four 20-inch pipes in

Khichri village, due to which the water level in this region went down by 10 ft.

The Coca-Cola plant in Plachimada was commissioned in March 2000 to produce 1,224,000 bottles of Coca-Cola, Fanta, Sprite, Limca, Thums Up, Kinley and Maaza. The local panchayat (council) issued a conditional licence for installing a motor to draw water, but the company started extracting millions of litres of clean water illegally from more than six borewells installed by it, using electric pumps in order to produce millions of bottles of soft drinks. According to the local people, Coca-Cola was extracting 1.5 million litres of water per day. The water level began falling, from 150 ft to 500 ft; as a result, the 260 borewells which had been provided by public authorities for drinking water and agriculture ran dry. Complaints were also received from tribals and farmers that sources of water and water storage were being adversely affected by the indiscriminate installation of borewells for tapping ground water, leading to serious consequences for crop cultivation in the area. Not only did Coca-Cola 'steal' the water of the local community, it also polluted what was left. In addition, the company was pumping waste water into dry borewells within the company's premises for disposing solid waste. During the rains, the solid waste spread into paddy fields, canals and wells, creating serious health hazards.

To add insult to injury, the company distributed the toxic waste from its plant as 'free fertiliser' to the villagers. Tests done on the waste showed that it contained extremely high levels of cadmium and lead, which can lead to cancer, kidney and liver disorders.

Although the local panchayat withdrew its license, the Kerala government gave Coca-Cola Rs 2 million in aid as part of its industrial policy. Such subsidies have been provided by every state government where a Coca-Cola or Pepsi plant exists. For local communities, every bottling plant is a source of the double hazard of man-made water scarcity and toxic waste dumping. Rural India is clearly a victim of the environmental and health costs of the soft drinks industry, but middle class urban India is also a victim because what Coca-Cola puts into the bottle is as toxic as what it leaves behind. The only difference is that the village women of Plachimada were aware of the threat posed to their health and survival, while affluent urban India is totally unaware of the harm soft drinks can cause. The Rs 6,247 crore spent annually by Indian consumers on soft drinks buys health hazards, not 'fun', as the ads promise.

When the local panchayat asked for details regarding the drawing of water, Coca-Cola failed to comply. The panchayat therefore served a show cause notice on them and cancelled the licence. Coca-Cola allegedly tried to bribe the panchayat president, A. Krishnan, with Rs 300 million, but he stood his ground. In 2003, the district medical officer informed the people of Plachimada that their water was unfit for drinking. The women already knew their water was toxic; instead of being able to draw water from the wells in their homes, they now had to walk miles for it. Coca-Cola had created water scarcity in a water-abundant region. The women of Plachimada were not going to allow this 'hydro-piracy'. They started a dharna at the gates of the Coca-Cola factory, and on September 21, 2003, a huge rally was

organised to give an ultimatum to Coca-Cola. A movement
started by local tribal women had created a national and
global wave of support. On February 17, 2004, the Kerala
Chief Minister, under pressure from the movement and the
aggravation of the water crisis because of a drought, ordered
the closure of the plant.

The victory of the movement in Plachimada was a
product of broad alliances, using multiple strategies. The
rainbow alliances, beginning with the local women and
activists like Veloor Swaminathan (convenor of the anti-
Coca-Cola task force in Plachimada) grew to include the
local village council (panchayat) and its members. Veerendra
Kumar of *Mathrubhumi* joined forces with it. The panchayat
used its constitutional rights to serve notice to Coca-Cola,
and the Perumatty panchayat filed a public interest litigation
in the Kerala High Court against the company. The courts
supported the women's demands. In an order passed on
December 16, 2003, Justice Balakrishnan Nair ordered
Coca-Cola to stop pirating Plachimada's water.

The World Water Conference in January 2004 was
co-organised with the local panchayat and brought together
every political party on one platform, and the leader of the
opposition, V.S. Achuthanandan, kept up the pressure in
the Kerala legislature to translate the court's decision into
executive action.

Kerala's active literacy movement provided leadership
through Dr Sukumar Azhikode,[2] and global support was
evident in the presence of European parliamentarians and
activists from across the world. The women's protest, the
heart and soul of the movement, gained support through

legal and parliamentary action and scientific research. This plural and diverse support was overwhelming and led to the victory of the people against Coca-Cola in Plachimada. I have always believed that immense power is unleashed when local agendas and global concerns unite to assert democratic access to nature's bounty.

Our movement to protect water and defend the water rights of all people and all species is called Jal Swaraj Abhiyan: the Water Democracy, Water Sovereignty Movement. It is part of our movement for Earth Democracy, Living Democracy, a movement in the defence of life, of the integrity and sanctity of the rivers, of our cultures, and of our capacity to be conservators and custodians of our precious and common water heritage. For us, water democracy is necessary for peace because centralised, commodified systems are creating water wars. We do not want blood for oil, nor do we want blood for water.

Notes

[1] Author of *Blue Covenant*, activist, and Right Livelihood Laureate. She is a founding member of the Council of Canadians, a citizens' advocacy organisation, and co-founder of the Blue Planet Project, which works internationally for the human right to water.

[2] Academic, orator, critic and writer of Malayalam literature, known for his contribution to Malayalam and insights on Indian philosophy.

No Patents on Life!

Biodiversity & Biotechnologies

MY LIFE'S JOURNEY HAS BEEN A JOURNEY OF BIODIVERSITY, of trying to understand how everything is interconnected through biodiversity – in the forests, on our farms, in our gut, our minds, our cultures and societies. Beginning with Chipko in the 1970s, and continuing with biodiversity in food and agriculture systems since 1984, my awareness and understanding about how the health of our planet and our health are connected through biodiversity have deepened and grown as COVID-19 and the health emergency engulf our lives today.

For over four decades I have worked on the reality of interconnectedness and non-separability; for me, protecting biodiversity is protecting both the integrity of life as well as the sovereignty rights, knowledge, and needs of local communities who have been the conservers and custodians of biodiversity. Over the years I have seen more clearly the limitations and violence of the mechanistic, militaristic, monoculture mind, which is why I have consciously cultivated a *biodiversity of the mind*, the ability to see life's processes in all their complexity and multiplicity. For me, knowledge and action, science and activism are one interconnected continuum.

Over the years, we have seeded diverse movements; a movement for the conservation of biodiversity and for saving seeds; a movement on 'No Patents on Life, No Patents on Seed'. What is described as the anti-GMO movement is also a movement to protect biodiversity and the integrity of life based on the deepest and highest science.

Prior to the Earth Summit in Rio in 1992, the Secretary in the Ministry of Environment in India invited me to advise them on the Convention on Biological Diversity (CBD) negotiations. When I inquired about who was leading the negotiations, he said it was the Director of Project Tiger. I told him that the biodiversity issue was no longer just about conserving big mammals, it was about saving every plant, every microbe. The Convention was the site of the biggest contest of our times. On one side were the big powers of the West who represented the interests of the emerging biotechnology industry that wanted to own, manipulate and exploit life through genetic engineering. I called this bio-imperialism. On the other side were the indigenous people and farmers, and nations of the South that had been biodiversity-rich but had been made economically poor by the extractivism of colonialism. I worked closely with the Third World Network (TWN) and like-minded governments to shape the Convention. Instead of allowing unregulated access to the biotechnology industry of the North, we ensured regulated access and the conservation of biodiversity.

We also built global movements for biodiversity, like Diverse Women for Diversity, and in Europe, the International Commission on the Future of Food and Agriculture (ICFFA), which I chaired, worked with local councils and governments to create GMO-free zones and regions. We signed an agreement and issued a joint declaration in November 2000.

In France, farmers like José Bové[1] uprooted GMO crops in Marseilles and destroyed GMO corn seeds. In Switzerland, the home of Syngenta (formed by the merger of Ciba-Geigy, Sandoz, Astra and Zeneca), the movement for a referendum on GMOs was successful: citizens voted against GMOs and the Swiss parliament has placed a moratorium on them which has been extended to 2025. Swiss biologist and chemist, Florianne Koechlin, a dear friend, played an important role in Switzerland becoming GMO-free.

The GMO-free movement is the reason why GMOs have not flooded Europe as they have America. In the US, I worked closely with Andrew Kimbrell of the Centre for Food Safety, and with Ronnie Cummins of the Organic Consumers Association. With activist and author Brian Tokar, we organised Biodevastation gatherings with citizens' participation. One of the most important gatherings was held in St Louis, Missouri, the headquarters of Monsanto. When we heard about Monsanto using the US government to sue Europe in the WTO for its moratorium on GMOs, we started a Citizens' Campaign on GMOs. More than 36 million people signed through their organisations and told the WTO that they did not want GMOs. Susan George,[2] José Bové and I presented the Citizens' Declaration at the WTO Hong Kong Ministerial in November 2005. The WTO did not rule against Europe.

I was also invited by local movements to launch the GMO-free campaign in South Africa. In Mexico, I supported the movement to prevent the entry of GMO corn, working with Adelita San Vicente Tello, who is now with the Ministry of Environment, and María Elena Álvarez-Buylla Roces, who heads the National Council of Science and Technology. Mexico has banned GMOs. In Brazil and Argentina, too, I have supported movements fighting Monsanto's GMO soya empire, and the Monsanto Tribunal in The Hague gave new strength to movements around the world.

*****

Biodiversity: A Third World perspective

The co-evolution of cultures, life forms and habitats has conserved biological diversity on this planet; cultural diversity and biological diversity go hand in hand. Communities around the world have developed deep knowledge and found ways to derive livelihoods from nature's diversity, in wild and domesticated forms. Hunting and gathering communities use thousands of plants and animals for food, medicine and shelter. Pastoral, peasant and fishing communities have also acquired knowledge and skills for sustainable livelihood diversity on the land and in rivers, lakes and seas. Today, however, the diversity of ecosystems, life forms and ways of life of different communities are under threat of extinction. Habitats have been enclosed or destroyed, diversity has been eroded, and livelihoods derived from biodiversity are endangered.

Tropical moist forests cover only seven per cent of the earth's land surface but contain at least half the earth's plant and animal species. Deforestation in these regions is continuing at a rapid pace, with very conservative estimates suggesting rates as high as 6.5 per cent in Côte d'Ivoire, and averaging about 0.6 per cent per year (about 7.3 million ha) in all tropical countries. In 1988, Peter Raven[3] estimated that about 48 per cent of the world's plant species occur in or around forest areas where more than 90 per cent of their habitat would be destroyed by the end of the 20th century. At the end of the century, he estimated that one million species would have been erased. Biological diversity in marine ecosystems is also remarkable, and coral reefs are sometimes compared to tropical forests in terms of diversity.

Livestock populations, too, have been homogenised and their diversity is being irreversibly lost. The carefully evolved pure breeds of cattle in India are on their way to extinction. The Sahiwal, Red Sindhi, Rathi, Tharpaker, Hariana, Ongole, Kankrej and Gir are cattle breeds developed for the different eco-niches where they survived and supported the needs of local communities. Today they are being systematically replaced by cross-breeds of Jersey and Holstein cows.

With animals disappearing as an essential component of farming systems and their contribution to organic fertility being substituted with chemical fertilisers, soil, fauna and flora are also on the brink of extinction. Locally-specific nitrogen-fixing bacteria and fungi that facilitate nutrient intake through mycorrhizal association, predators of pests, pollinators and seed dispersers, and other species that co-evolved over the centuries to provide environmental services to traditional agro systems have vanished, or have seen their genetic base shrink dramatically. Deprived of the flora with which they co-evolved, soil microbes have also disappeared.

Biodiversity erosion starts a chain reaction. The disappearance of a species is related to the extinction of innumerable other species with which it is inter-related through food webs and food chains. The crisis of biodiversity is not just a crisis of the disappearance of species that have the potential of spinning dollars for corporate enterprises; it is basically a crisis that threatens the life-support systems and livelihoods of millions of people everywhere.

There are two primary causes for the large-scale destruction of biodiversity. The first is habitat destruction due to internationally financed mega-projects such as dams, highways, and mining operations in forested regions. The second is the technological and economic push to replace diversity with homogeneity in forestry, agriculture, fisheries and animal husbandry. The Green Revolution in agriculture, the White Revolution in dairying and the Blue Revolution in fisheries are 'revolutions' based on the deliberate replacement of biological diversity with biological uniformity and monocultures.

In Thailand, the Nam Choan Dam flooded the valley of the Thung Yai and Huai Khaeng Wildlife Sanctuaries, which together comprise the largest intact block of forested land set aside for wildlife conservation in the country. The dam threatened the habitat of the largest remaining populations of elephant and banteng, as well as a variety of other threatened or endangered species such as tiger, tapir and birds like the green peafowl.

In Brazil, the Grande Carajás Programme involving the Tucurui dams, as well as the iron ore and bauxite mining and processing industry, have seriously impacted biodiversity and cultural diversity in the Amazon. Amazonia harbours more wildlife than any other place on earth, both per unit area and as a subcontinental region. There are estimated to be over 50,000 species or higher, and at least an equal number of fungi; one-fifth of all the birds on our planet; at least 3,000 species of fish, amounting to ten times the number of fish in all the rivers of Europe; and insect species numbering in the uncounted millions. The great age and size of the forests, their favourable climate (hot and moist), the fact that they have remained undisturbed for millennia, and the presence of very high concentrations of species in particular areas have all contributed to the region's unparalleled diversity.

During the filling of the Tucurui reservoir, which flooded over 3,000 square kilometres of rainforest over twenty years, an attempt was made to save the drowning animals. In one ten-day period, 4,037 mammals, 4,848 reptiles, 6,293 insects such as giant scorpions and spiders, 717 birds and 30 amphibians were captured by men in boats – some 15,925 creatures from one part of the lagoon. Brazilian ecologists estimated that this total was only a tiny fraction of the actual number held by the forest.

The two principles on which the production and maintenance of life are based are: diversity; and symbiosis and reciprocity, also called the law of return. The two are not independent; they are interrelated. Diversity gives rise to ecological space for give and take, for mutuality and reciprocity. Destruction of diversity, of the self-regulated and decentralised organisation of diverse systems, gives way to external inputs and centralised control. Sustainability and diversity are ecologically linked because diversity offers the multiplicity of interaction which can heal ecological disturbance to any part of the system.

In spite of the immeasurable contribution that biodiversity in the South has made to the wealth of industrialised countries, corporations, governments, and aid agencies in the North continue to create legal and political frameworks to make the South pay for what it originally gave to the North. Emerging trends in global trade and technology work inherently against justice and ecological sustainability and threaten to create a new era of bio-imperialism, built on the biological impoverishment of the biosphere.

In the self-provisioning economies of the South, producers have traditionally been conservers. It has been recognised that the total genetic change achieved by farmers over the millennia has been far greater than that achieved by the one hundred or two hundred years of more systematic science-based efforts. If this contribution to knowledge and the development of biodiversity is recognised, farmers and tribals become the original producers, and corporate and public sector scientists consume their finished products as raw material for commodities. The dominant approach turns this relationship of producer and consumer on its head.

An economistic bias narrows down conservation options to commercialised approaches in which both the means and ends of conservation are financial values on the market. Commercialised conservation is linked to the emergence of new biotechnologies, which have transformed the genetic resource of this planet into raw material for the industrial production of food, pharmaceuticals, fibres, energy, and so on. Commercialised conservation justifies the value of conservation in terms of its present or future use for profit. Biodiversity conservation here is assessed only in terms of setting aside 'reserves' in undisturbed ecosystems for the purpose of conservation. This schizophrenic approach to biodiversity, which adopts a policy of destruction of diversity in production processes, and a policy of preservation in 'set-asides' cannot be effective in the conservation of species diversity. Biodiversity cannot be conserved unless production itself is based on a policy of preserving diversity.

The dominant approach to biodiversity is inadequate for conservation both because it values biodiversity only as a commodity, and also because it thinks of biodiversity as fragmented and atomised, merely as arithmetic, a numerical, additive category. Ex situ conservation in high-tech gene banks has been the main response to biodiversity conservation, but this approach is both static and centralised. It may be an efficient means of conserving raw material in the form of germ plasma collection, but it has serious limitations because it not only removes control over biodiversity from local communities, it also removes biodiversity from the habitats where diversity can evolve and adapt under changing environmental conditions.

I have always held that an ecologically sustainable and just approach to biodiversity conservation needs to begin by halting and reversing the primary threat to biodiversity. This involves stopping aid and incentives for the large-scale destruction of habitats where biodiversity thrives, as well as subsidies and public support for the displacement of diversity by centralised and homogeneous systems of production in forestry, agriculture, fisheries and animal husbandry.

Ecologically, this shift involves recognising the value of diversity in itself. All life forms have an inherent right to life, and that should be the overriding reason for not allowing species extinction to take place. At the social level, the values of biodiversity in different cultural contexts need to be recognised and conserved. In addition, we need to understand that market and dollar values are very limited values; there are other values of biodiversity, such as providing meaning and sustenance, and these values should not be treated as subservient and secondary to market values.

Ecology, equity and efficiency meet in biodiversity, while they are in opposition with each other in monocultural and homogeneous systems. Neither ecological sustainability nor livelihood sustainability can be ensured without a just resolution of the issue of who controls biodiversity. After centuries of the gene-rich South having contributed biological resources freely to the North, governments in the South are no longer willing to have biological wealth appropriated for free and sold back at exorbitant prices to them, as 'improved' seeds and packaged drugs.

To redress the North-South imbalance and to recognise the contributions of local communities to the development of biodiversity, it is imperative that regimes based on bio-imperialism be replaced by structures based on bio-democracy. Bio-democracy involves recognising the intrinsic value of all life forms and their inherent right to exist. Pro-democracy entails that nation-states protect these prior rights from erosion by corporate claims to private property in life forms through patents. The deeper the devolution and decentralisation of rights to biodiversity, the fewer the chances for monopolising tendencies to take hold.

*****

The fact that one of the agenda items for the UN Conference on Environment and Development (UNSD) in 1992 was the 'environmental management of biotechnology' indicates that biotechnology is surrounded by social and ecological anxiety.

The first anxiety arises from the fact that the new biotechnologies tamper with the very fabric of life, and demand a fundamental restructuring of our minds, our ethics, our environment, social and economic values and relationships. While biotechnology in its broadest sense is a very ancient group of technologies, it is the new technologies that generate new social, ecological, economic and political risks.

The new biotechnologies consist of two major groups: the first group, genetic engineering, refers to the new techniques derived from advances in molecular biology, biochemistry and genetics. The second group is based on new cellular procedures that draw on the older technology of tissue culture.

Genetic engineering is a powerful technique which theoretically allows any gene to be moved from any one organism into another. Recombinant DNA technology has the potential to transform genes into a global resource that can be used to shape novel life-forms. It is this technical power which gives it the potential to become more pervasive than any other biotechnology. Past biotechnology had already found application in primary industries (agriculture, forestry and mining); in secondary industries (chemical, drugs and food); and in tertiary industries (healthcare, education, research and advisory services).

In addition to the wide-ranging application of biotechnology is the fact that the development of new technologies is almost entirely controlled by transnational corporations, even though universities and small firms evolved the techniques. These corporations are diversifying into every field of specialty which uses living organisms as a means of production. Traditional industry sectors are becoming less distinct, and corporate boundaries virtually unlimited. This integration, centralisation and control carries an inherent destabilisation at the socio-economic level.

Technological innovation and scientific change do not merely bring benefits, they also carry social and ecological costs. It was the scientists closest to genetic engineering who first expressed concerns relating to the emergence of the new technology. In 1973, a group of prominent scientists called for a moratorium on certain types of research due to unknown risks and hazards associated with the possible escape and proliferation of novel forms of life. In 1975, at the Asilomar Conference in California, part of the scientific community led by Paul Berg, a molecular biologist at Berkeley, attempted to agree on the need for regulation of biotechnological research.

Later, as many scientists got involved in the commercial application of these new technologies – what Congressman Al Gore called the 'selling of the tree of knowledge to Wall Street' – the self-criticism and self-restraint within the scientific community receded.

Sustaining the social impact analysis of new technologies then became the responsibility of individual scientists and activists. The most persistent theme in their critique has been the fear of adverse ecological and epidemiological consequences that might stem from the accidental or deliberate release of self-propagating genetically engineered organisms into the biosphere. Prominent scientists like Liebe Cavalieri, George Wald and David Suzuki have argued that the very power of the new technologies outstrips our capacity to use them in safety, that neither nature's resilience nor our own social institutions are adequate protection against the unanticipated impacts of genetic engineering.

This brings me to the 'ice-minus' story and the public outcry against its testing and deliberate release in the North. Since frost damage is a major threat in the cold climates of the North and runs up US $14 billion in cost annually, biotechnologists worldwide have tried making plants more tolerant to frost. They isolated a gene which triggers ice nucleation in a plant cell, and deleted it from a certain bacterium called pseudomonas syringae. The idea was that when the ice-minus bacteria are sprayed on a crop, such as Californian strawberries, they displace the naturally occurring ice-forming bacteria and prevent the plants from freezing as they would normally.

In 1983, Steven Lindow[4] of Berkeley, and Advanced Genetic Science, a firm funding his work, were permitted by the National Institutes of Health's (NIH) Recombinant DNA Advisory Committee to run a field test for this new technology. However, on September 4, a group of citizens and environmentalists based in Washington, DC – including Jeremy Rifkin and the Foundation on Economic Trends, the Environmental Task Force, Environmental Action, and the Human Society – filed a public interest suit against the

NIH for approving the project. Among other things, the suit charged that the NIH had not conducted an adequate assessment of the potential environmental risks of Lindow's field test and had 'been grossly negligent in its decision to authorise the deliberate release of the first genetically engineered life-forms'.

Among the risks that the suit mentioned was the dramatic possibility that the frost-preventing bacteria might be swept into the upper atmosphere, disrupting the natural formation of ice-crystals, ultimately affecting local weather patterns and possibly altering the global climate. Eminent scientists like Eugene Odum[5] and Peter Raven pointed to the ecological hazards of deliberately releasing micro-organisms since they reproduce rapidly. Moreover, their inter-relations with higher plants such as trees are not known.

The public outcry associated with the ice-minus field test pushed Northern governments and corporations into taking their trials overseas to countries with little or no regulation, which meant the Global South.

As bans and regulations delay tests and marketing in the North, biotechnology products have increasingly been tested in the South so as to bypass regulation and public control. The public, scientists, and official agencies in countries where these technologies are being developed are aware of these hazards. Genetic engineering, therefore, faces regulatory constraints, public protests and injunctions, domestically; consequently, experiments involving recombinant organisms are conducted in countries where obstacles appear to be fewer due to lax legislation and lower public awareness. The Indian government, for example, welcomed the biotech bandwagon of foreign companies by diluting regulations and eroding the democratic structures that have existed within the country. The Vaccine Application Program (VAP), for example, is clearly designed to bypass safety regulations prevalent in the US, because the Memorandum of Understanding states that all genetic engineering research 'will be carried out in accordance with the laws and regulations of the country in which the research is conducted'. Since India has no laws or regulations regarding genetic engineering, testing vaccines in India amounts to totally unregulated, deliberate experimentation.

The VAP was initiated in 1985 as part of the Reagan-Gandhi Science and Technology Initiative, and the agreement was signed in Delhi on July 9, 1987. The Program document stated that

> The announcement of the VAP is an important recognition that vaccines are among the most cost-effective of health technologies, and their widespread use in both countries is key to controlling the burden of vaccine-preventable disease.

The primary purpose of the VAP was allowing an extended range of trials of bioengineered vaccines on animals and human subjects. The priority areas were identified as typhoid fever, rotavirus, hepatitis, dysentery, rabies, pertussis, pneumonia and malaria, but these could change in succeeding years as other areas of research opportunity were identified.

In 1986, the Wistar Institute in Philadelphia hit the headlines for testing bioengineered rabies vaccines on cattle in Argentina without the consent of the government or the people. When the government became aware of the bovine rabies vaccine experiment in September 1986, it was immediately terminated. The Argentinian minister of health alleged that farmhands who cared for the vaccinated cattle had been infected with the live vaccine. Wistar was driven out by the Argentinian government, but was welcomed by the Indian government for participation in the VAP. In fact, the project paper for the VAP, prepared by the US government, applauded Wistar for its achievements in the field of vaccine development, and specifically mentioned the bovine rabies vaccine for field trials and other research.

The VAP was financed by United States Agency for International Development (USAID) and US Public Service. The total project cost was US $9.6 million, of which the US and Indian components were US $7.6 million and US $2 million, respectively. Through its financial input the US government controlled the agreement; thus all 'documents, plans, specifications, contracts, schedules and other arrangement with any modifications therein, must be approved by USAID'. On the other hand, scientists and scientific agencies in India directly concerned with the subject were excluded from the discussions.

The controversial Indo-US vaccine project bypassed the high-powered biotechnology scientific advisory committee set up by the Government of India. Dr Pushpa Bhargava, a member of the committee and director of the Centre for Cellular and Molecular Biology, said that the steps postulated in the vaccine agreement 'are bound to come in the way of setting up our own research and development, and threaten to compromise our national sovereignty'. Then Union science minister, K.R. Narayanan, was not informed of the details of the agreement, nor was Dr V.S. Arunachalam, science advisor to the defence minister. The director general of the Indian Council of Medical Research (ICMR) stated categorically that he would not allow any vaccine to be tried on Indians unless the same was also approved for use in the US. As a result of scientific and public protest, the implementation of the VAP became even more secretive and, gradually, totally removed from the public gaze.

A programme that would expose the Indian public to the known hazards of live viruses used as vaccines denies the human subjects of the vaccine experiment the right to prior informed consent. Human beings everywhere have a fundamental right to be told when they are being treated as guinea pigs, and they have the right to refuse to participate if they fear exposure to unnecessary risk. With genetically engineered vaccines, the risks are indeed very high. Most researchers consider the use of attenuated lethal viruses as live vaccines too risky; creating hybrid viruses has been viewed as one way to circumvent these risks. Recombinant DNA technology can be used to add the gene for an antigen of a lethal virus to the genome of a harmless virus, in an attempt to create a harmless living hybrid virus which, if used as a vaccine, provides immunity against the lethal virus. However, as Peter Wheale and Ruth McNally report in *Genetic Engineering: Catastrophe or Utopia?*, research has shown that genetic manipulation of harmless viruses can render them virulent. There is no 'safe' bioengineered vaccine. While the VAP is totally irresponsible with regard to the protection of people's health and environmental safety in light of this hazardous implication, it shows great concern for the protection of corporate profits. It has a special clause for an accord on content, on intellectual property, which attempts to undo the public interest content of the Indian patent protection system.

Argentina and India are not the only countries to which biohazards were being exported. At a week-long National Conference on Plant and Animal Biotechnology, held in Nairobi in February 1990, USAID officials were pressing African states to allow field trials of genetically-altered organisms that might not have been allowed by regulatory systems in the North. Such was the concern that the minister for research, science and technology made a public pledge on the conference's second day, stating that Kenya would not become a testing ground for dangerous new biotechnology products. Dr Calestous Juma, Director of the Nairobi-based African Centre for Technology Studies (ACTS), advised scientists that USAID was encouraging countries in Asia and Latin America to undertake similar testing roles for private American firms.

Hazardous substances and processes have been manufactured faster than structures of regulation and public control have evolved. We do not yet have full ecological criteria for testing an environmentally safe management of the fossil fuel technologies of the mechanical engineering revolution; the tests for an environmentally safe management of the chemical engineering revolution are still in their infancy, leading to the marketing of products, processes and waste which are proving to be ecologically unmanageable. Tests for safety in the genetic engineering revolution are yet to be conceived, since how genetically modified life-forms interact with other organisms is an unknown and uncharted territory.

In biotechnology, more than in any other area, the lack of knowledge about hazards cannot be treated as 'safety'. Restraint and caution are, therefore, considered the only wise strategy for unleashing powerful technologies with potentially serious risks in a context of near-total ignorance.

For countries of the South, a special danger exists in being used as a testing ground and as guinea pigs. In addition, uncertainties are aggravated by the fact that the governments of the South want access to the new technologies of the North; in their haste to acquire new biotechnologies they could unwittingly find themselves, and their people and the environment, becoming the testing ground.

So, in order to increase the benefits of new technologies and reduce their negative impact, the South needs to rapidly evolve a framework for assessing biotechnology on the basis of its ecological, social and economic impact. Transfer of technology, an important issue for the South, needs to be negotiated within such an assessment framework, so that a socially desirable technology transfer can take place and undesirable and hazardous transfers can be prevented.

There is a prevalent misconception that biotechnology development will automatically lead to biodiversity conservation. The main problem with viewing biotechnology as a miracle solution to the biodiversity crisis is related to the fact that biotechnologies are, in essence, technologies for the breeding of uniformity in plants and animals. Biotech corporations talk of contributing to 'genetic diversity'; John Duesing of Ciba-Geigy (now Novartis) stated, 'Patent protection will serve to stimulate the development of competing and diverse genetic solutions, with access to these solutions ensured by free market forces at work in biotech ecology and seed industries.' However, the 'diversity' of corporate strategies and the diversity of life forms on this planet are not the same thing, and corporate competition can hardly be treated as a substitute for nature's evolution in the creation of genetic diversity.

Corporate strategies and products can lead to diversification of commodities; they cannot enrich nature's diversity. This confusion between *commodity diversification* and *biodiversity conservation* finds its parallel in raw material diversification. Although breeders draw genetic materials from many places as raw material input, the seed commodity that is sold back to farmers is characterised by uniformity. Uniformity and monopolistic seed supplies go hand in hand. When this monopolising control is achieved through the molecular mind, the destruction of diversity is accelerated. As Jack Kloppenburg[6] warned, 'Though the capacity to move genetic material between species is a means for introducing additional variation, it is also a means for engineering genetic uniformity across species.'

Perhaps the most serious impact of biotechnology is the displacement of some agricultural export commodities from the South with related impacts on national economies and employment. Plant tissue culture offers increased possibilities for substituting specialties with industrially produced inputs. Many high value plant-derived products used for pharmaceuticals, dyes, flavourings and fragrances are vulnerable to displacement as a result of current research.

The impact of the successful production of substitutes will be felt most by countries which have, in an earlier international division of labour, been made dependent upon exporting the natural products concerned. This proved to be particularly destructive to economies in Africa, which depended entirely on single crops for most of their export earnings. While historically Africa has grown crops needed for Europe, in a world order based on new biotechnologies, Africa became dispensable as the North began to find biotech substitutes for African crops.

When factories close in the North, compensation is paid to workers. When crops introduced by global agribusiness are displaced by the technologies developed by agribusiness, small peasants and agricultural workers are left to fend for themselves, as are their countries. The South needs to develop an agenda for compensation which is based on historical justice, and which can be tabled before the full deployment of the new biotechnologies that are being developed to reduce dependence on the South takes place.

*****

Most of the adverse impacts of biotechnology are related to the fact that the new technologies are evolving under the control of the transnational private sector. Biotechnology was born in the laboratories of universities and other public research institutions. Some scientists then moved out to set up their own biotech companies. Now it is giant agrochemical, pharmaceutical and food processing transnationals that dominate both the research and the markets.

Along with this trend in privatisation is the trend towards concentration. Where, in the mid-1970s, there were thirty manufacturers involved in pesticide development in the US, there are only a dozen today. For decades, the top thirty drug producers remained the same; today, ten corporations control 28 per cent of the world drug market, as a result of mergers.

Transnational corporations have bought up most of the seed companies; by 2000, the top ten companies controlled most of the seed market, including that owned by farmers who save their own seed, as well as that controlled by the public agriculture research system which played a central role in the development and distribution of Green Revolution seed varieties.

Privatisation trends are also indicated in China, a pioneer in developing hybrid rice which is capable of boosting harvests by 25 per cent. But the rice variety that allows its production – a so-called male-sterile line of rice that will not self-seed – is not distributed in Asia. Two multinational companies – Cargill Seeds and Occidental Petroleum's Ring Around Products – are known to have entered into exclusive licensing agreements with the Chinese government for seed development, production and marketing in specified countries.

An agreement between the Chinese government and the two US companies forbade the sharing of information and materials concerning hybrid rice with other governments or with the International Rice Research Institute (IRRI); the Chinese government was therefore forced to withdraw its supports for an IRRI hybrid rice training course.

The divergence between the imperative for private profit and people's well-being is only expected to grow – corporations will attempt to adjust society to their need for profit. They will increasingly use the state to restructure relationships between people, and between North and South, to further their interests. Privatisation is an increasing threat to democracy and people's will, as the same scientists who work on contract for transnational corporations function on government regulatory bodies and dominate scientific research. In this context, it is up to citizens, free of TNC and government control, to keep public issues and priorities alive, and create the space for public control of new biotechnologies.

The ultimate expression of privatisation of biotechnology is the desperate urge by TNCs, operating through the US Trade Representative, World Bank, GATT/WTO, to ensure a uniform patent system across the world that allows them to appropriate all life on this planet as their private property.

Patents in the context of agriculture and food production involve ownership over life forms and life processes. A monopoly ownership of life creates an unprecedented crisis for agricultural and food security, by transforming biological resources from commons into commodities. It also generates a crisis of the values and ends which guide social organisation, technological change and development priorities. Yet, it is not the countries that are demanding worldwide intellectual property protection that are benefitting from it; it is multinational corporations. Monsanto's Nicholas Reding says, 'The major challenge to genetic engineering scientists and companies, as well as national governments, is to support uniform worldwide property rights.' This is just another way of saying that global monopoly over agriculture and food systems should be handed over as a right to multinational corporations. With wide patent protection, agribusiness and the seed trade are trying to achieve a truly global reach. While the rhetoric is 'agricultural development' in the South, the enforcement of strong patent protection for monopoly ownership of life processes undermines agriculture in the South in a number of ways.

The corporate demand to change a common heritage into a commodity and to treat profits generated through this transformation as a property *right* will lead to erosion not just at the ethical and cultural, but also at the economic level for farmers in the South, who have a three-tier relationship with corporations that demand a monopoly over life forms and life processes. First, s/he is a *supplier* of germ plasm to TNCs; second, s/he is a *competitor* in terms of innovation and rights to genetic resources; third, s/he is a *consumer* of the technological and industrial products made by TNCs. Patent protection displaces farmers as competitors, transforms them into suppliers of free raw material, and makes them totally dependent on industrial supplies for inputs like seed. Above all, the frantic cry for patent protection in agriculture is *protection from farmers*, the original breeders and developers of biodiversity resources in agriculture.

Unlike plant breeders' rights (PBR), the new utility patents are broadbased, enabling monopoly rights over individual genes and even characteristics. PBR is not an ownership over germ plasm in seeds; it grants a monopoly right only for the selling and marketing of a specific variety. The monopoly rights of industrial patents go much further. They allow for multiple claims not only on whole plants but on plant parts and processes as well. So, according to attorney Anthony Diepenbrock,

You could file for protection of a few varieties of crops; their macro parts (flowers, fruit, seeds, and so on); their micro parts (cells, genes, plasmids, and the like); and whatever novel processes you develop to work with these parts, all using one multiple claim.

Patents are the hurdles that remain to be crossed for the large-scale distribution of biotech seeds by TNCs. One of the clauses in India's new seed policy in 1989 directed all companies importing seeds to make a small quantity available to the gene bank of the government-controlled National Bureau of Plant Genetic Resources (NBPGR). Corporate giants were, of course, unwilling to accept that clause and wanted it to be removed, which it was. As Jan Nefkins, general manager of Cargill South-East Asia Limited, pointed out, 'No company would be willing to part with what they took years, and spent millions of dollars, developing. It's a question of intellectual property rights.'

One outcome of property rights over living systems has been secrecy in plant breeding and plant genetics research, and restrictions on the exchange of germ plants. Secrecy for patents and exclusivity together choke all scientific exchange in plant genetics. In addition, rather than simply stimulating innovation, the patent system as applied to living matter redirects attention towards those products that provide for the broadest and easiest patent protection, not towards those for the larger public good.

India's patent laws have excluded the monopoly over biological processes which are essential to sustenance and survival. Food, plants and animals have been excluded from patentability. However, living organisms are central to production processes in biotechnology; the need for ownership rights to living organisms is essential for the next stage of capital accumulation by global corporations, and at the core of their business are patents which guarantee profits.

Who owes whom is a tricky issue, especially when it comes to profiteering from biological resources which have originated in the South and continue to provide sustenance and survival to millions of farmers. The Indian patent law has excluded private ownership of the biological foundations of agriculture to ensure that entitlements to food and nutrition are as broadbased as possible. Sovereignty in the matter of patent law is essential because it is a matter of survival, especially for economically weaker sections in our society that have no purchasing power and can be protected only through public interest. The choice is clear: protection of life vs. the protection of profits, a choice that the world witnessed in its starkest manifestation during COVID – pre-tax profits of Pfizer-BioNTech and Moderna were estimated to be $34 billion in 2020-21. Their refusal to share their mRNA technology with other pharma companies has deprived the poor of the world protection against the virus at an affordable price.

*****

Over the last few decades we have questioned the dead earth paradigm imposed by dominant corporations. We have placed biodiversity at the crux of our perception of, and our relationships with, the living earth and in our role as regenerators of her biodiversity. Nature does not work on the principle of sameness, uniformity or monocultures; the natural world is a constant striving for diversity of expression. Cultures, too, seek diversity, and cultural diversity flows from nature's principles and her biodiversity.

Diverse ecosystems give rise to diverse life forms and diverse cultures. The co-evolution of cultures, life forms and habitats has created, regenerated, and conserved biological diversity and cultural diversity on this planet. Cultural diversity thrives when societies and communities are free to look after their ecosystems and resources, share them in the commons, and use them sustainably for the common good.

We have rejuvenated forgotten knowledges and forgotten foods, and through our ideas and actions shifted the ground from the perception of biodiversity as inert and valueless raw material to the recognition that it is the organising principle of life, the foundation of living economies and living cultures. Biodiversity weaves the fabric of the natural and social worlds.

Diversity is the basis of ecological stability in nature and social stability in societies. Sustainability and diversity are ecologically linked, because diversity offers the multiplicity of interaction which can heal ecological wounds to any part of the system in the natural and social worlds of which we, too, are a part.

The past three decades have also shown how, instead of being a miracle technology to control weeds and pests, biotechnology and GMOs are a failed intervention. New

GMOs are being introduced to cover up the failure of old GMOs, the failure of Bt cotton to control pests and the failure of Round-Up Ready crops to control weeds. They are also aimed at creating a new narrative for genetic engineering while maintaining the genetic determinism and genetic reductionism paradigm on which genetic engineering is based. They extend the illusion of mechanistic determinism in an attempt to 'engineer' complex, living, self-organised systems. They are also an attempt to escape regulation for their impact on biodiversity and health.

New GMOs based on gene editing (CRISPR – Clustered Regularly-Interspaced Short Palindromic Repeats, Cas9) are being rushed to the market in spite of uncertainties related to the technology. They are being sold as natural, in an attempt to bypass biosafety regulation. There is a false claim being made that CRISPR is not a GMO, but modifying an organism at the genetic level is a scrambling of the self-organisation and intelligence of the living organism, which has unpredictable outcomes. And just as old GMOs were based on the fiction of bringing miracles to farmers, the biotechnology industry, based on new GMOs, is propagating the fiction of bringing benefits to consumers through 'natural', 'healthy' foods based on CRISPR.

In the 1980s, Monsanto was in the forefront of the push for GMOs and patents on life. Today it is Bill Gates. One rich individual is able to use his wealth to bypass all international treaties and all multilateral governance structures to, in turn, enable global corporations to grab the fruits of biodiversity, to control farmers' seeds, and violate national and international laws that protect farmers' rights and seed sovereignty.

Bypassing international treaties that protect seed sovereignty is the clear aim of the digital colonisation of seed. The International Digital Council for Food and Agriculture proposed data gathering on a global scale at the Global Forum for Food and Agriculture in Berlin in 2019. It said,

> These practices would extend to genetic digital sequence information (DSI or digital DNA) and could preempt already-contentious access and benefit sharing (ABS) negotiations in the Biodiversity Convention and the FAO Seed Treaty.

This is why digital colonisers say that data is the new gold.

The biotechnology industry, the technology giants, and financial giants like BlackRock are now joining hands to financialise and monetise biodiversity. In mid-October 2021, before the start of the Conference of Parties to the Convention on Biodiversity, COP26, Wall Street launched a new class of financial assets on all elements of nature, including biodiversity. Asset Management companies like BlackRock and Vanguard – of which BlackRock is the main shareholder – are looking to reduce the natural world to financial capital. Indeed, BlackRock has estimated the worth of our natural world to be four quadrillion dollars, or 4,000 trillion dollars. Consider that the world GDP in 2020 was valued at 84.5 trillion dollars. This, then, is expected to be the biggest land grab, biodiversity grab, wealth grab, in history. To stop the colonisation of nature and our cultures, our movements have to become more imaginative, more resilient, more unified. For the colonisation of life will extinguish the freedom to live for all life and all people on the earth.

Notes

[1] French farmer, politician, and syndicalist, and spokesman for Via Campesina.

[2] Political and social scientist, activist and writer on global social justice, Third World poverty, underdevelopment and debt. She is the president of the Transnational Institute, a think-tank located in Amsterdam, and a fierce critic of the 'maldevelopment model' of the IMF and World Bank.

[3] American botanist, environmentalist and President Emeritus of the Missouri Botanical Garden.

[4] Plant pathologist and microbial ecologist, Lindow is Professor Emeritus, Department of Plant and Microbial Biology, University of California, Berkeley.

[5] American biologist known for his pioneering work on ecosystem ecology.

[6] Professor Emeritus in the Department of Community and Environmental Sociology at the University of Wisconsin-Madison, he is well known for his analysis of the emergent social impact of biotechnology and for his work on the global controversy over access to, and control of, biodiversity.

Staying Alive

Climate Chaos/
Climate Action

THE CONVENTION ON BIODIVERSITY, AN INTERNATIONAL treaty to prevent biodiversity erosion and regulate emissions of greenhouse gases that are contributing to climate change, and the United Nations Framework Convention on Climate Change were signed simultaneously in 1992 at the Earth Summit in Rio. However, over the years, the two environment issues were delinked from each other, with the result that the biodiversity crisis and climate crisis deepened and have now become emergencies. Meetings of the Conference of the Parties have failed to address them; instead, they have become a platform for transnational corporations and the world's rich, who are mainly responsible for both climate change and biodiversity destruction, and who now want to find new ways to control and commodify nature in the service of profit.

In the lead-up to the Climate Change Conference in Copenhagen in 2009, I wrote *Soil Not Oil* in order to reconnect biodiversity and climate change. My book grew out of a talk I gave for the Transition Town Movement in Bristol, organised by Patrick Holden,[1] then director of Soil Association and Rob Hopkins, founder of Transition Town, a grassroots community initiative in over 50 countries that seeks to increase self-sufficiency in order to reduce the potential effects of peak oil, climate change and economic instability. There are now close to 2,000-3,000 communities in Transition Town initiatives across the world. The challenge that this quiet networked revolution wanted to address was how we can change agriculture and food systems so as to reduce greenhouse gases and make agriculture and food systems more climate resilient.

I have witnessed how oil has devastated the land and communities. Shell has destroyed the Niger delta, and polluted its rivers and villages. Ken Saro-Wiwa, a leading Nigerian writer and environmental activist, took them on in the 1990s, and Shell had Ken arrested. On November 10, 1995, Ken was hanged on false charges. I remember we were at the Riverside Church in New York for a teach-in organised by the International Forum on Globalisation (IFG) when the news of Ken's death reached us.

Twenty years later I was invited by Nnimmo Bassey and the Right Livelihood College in Port Harcourt. My dear friends, the late Anita Roddick and her husband, Gordon, who founded The Body Shop, were big supporters of the people's resistance to Shell and Big Oil. When Anita passed away on September 10, 2007, Gordon organised a memorial for her and invited me to speak at it. No politicians were allowed to talk. In her last interview with the BBC, when Anita was asked whether she wanted to be remembered as the founder of The Body Shop, she said, 'No, I want to be remembered for being an activist.' So after the memorial, we marched to the Shell office in London with placards saying 'I am an activist.'

The Government of Ecuador invited me to be an Ambassador for the Yasuni National Park in the Amazon, as well as an ambassador to make visible the crimes of the oil giant, Chevron, against indigenous people. When the British Petroleum Deepwater Horizon oil spill took place in the Gulf of Mexico, I was invited by different environment movements to file a case in the Constitutional Court against BP. Ecuador's Constitution is based on the Rights of Nature. The Ecuadorian Ambassador to India brought me an invitation from the President, Rafael Correra, when the new Constitution was pronounced and he was elected to office. The Ambassador said the President had been deeply inspired by my book *Staying Alive*. On November 2, 2010, a case was filed against BP by Nnimmo Bassey, myself, and others on the grounds of Rights of the Earth.

As I wrote in *Soil Not Oil*, the transition for climate action is a transition from oil-based thinking and living to soil-based thinking and living. How we grow, process, and distribute our daily bread is the single biggest contributor to greenhouse gases that are driving the climate crisis by polluting the atmosphere and disrupting the earth's and nature's self-regulating ecological systems.

At the Copenhagen Climate Change Conference, the International Commission on the Future of Food and Agriculture (ICFFA) also presented the *Manifesto on Climate Change and the Future of Food Security*, prepared by Shumei International of Japan to more than one thousand climate activists. I organised a panel on how ecological, organic agriculture is a climate solution. Maurice Strong of the UN Environment Programme, who chaired the 1992 Earth Summit, was also present, as was Patrick Holden of the Soil Association. After Copenhagen, the movement connecting soil and agriculture started to grow. The official Copenhagen Climate Change Conference was derailed by US President Barack Obama when he joined hands with the principal polluters to undertake voluntary commitments, instead of endorsing the legally binding emissions reduction targets that are at the crux of the UN Framework Convention on Climate Change. That is when President Evo Morales of Bolivia said, 'We are here to protect the rights of mother earth, not the rights of polluters.' He organised a World People's Conference on Climate Change and the Rights of Mother Earth, which then led to the Universal Declaration of the Rights of Mother Earth.

At the UN General Assembly in September 2009, the UN voted to refer to April 22, Earth Day, as Mother Earth Day. The Declaration was presented at the UN General Assembly on April 22, 2020, on 'Harmony with Nature', where I was invited to speak.

The Global Alliance for the Rights of Nature, of which I am a founding member, also grew out of the Declaration of the Rights of Mother Earth. The failure of Copenhagen created a new opening for citizens and governments to start seeing the earth as alive, moving beyond the anthropocentric and corporocentric bias of looking at nature as inert and passive.

I wrote *Oneness vs. the 1%* with my son, Kartikey, as I watched the 2015 Paris Climate Change Conference being hijacked by Bill Gates. COP26 in Glasgow, in November 2021, further accelerated the capture of the UN system by Big Business and their false solutions.

The climate crisis is the result of a two-century error made by the colonising industrial world in making production, trade and consumption dependent on fossil fuels. Nature created coal and oil by fossilising the living carbon of plants and other organisms over 600 million years. She placed it underground for us to leave it there. Instead, we hurtled towards destroying the earth, her forests and farms, soil and water, climate systems and biodiversity, chasing the illusion that we were on the road to progress.

Over the last century, beginning with the global monopoly of Standard Oil, the world was forced into dependence on it, a dependence that was defined as 'development'. Climate Action represents a de-addiction from fossil fuels. Fossil fuel addiction has created a mechanistic way of thinking that I call a 'fossilised' mindset, an industrial mode of production for meeting our daily needs. We eat oil. We drink oil. We breathe oil. The fossil fuel age has created government policies and economic strategies which privilege oil and oil-based systems, punishing soil and soil-based local, living economies.

Fossil fuel-based industrialism displaces and destroys biodiversity and contributes to climate change. Climate disruption intensifies droughts, floods and cyclones. Extreme events have a more disastrous impact when the cushion of biodiversity has been destroyed.

Industrial corporate agriculture, based on fossil fuel, chemical-intensive agricultural production, and the globalised corporate industrial food systems based on long distance transport and food miles; energy-intensive ultra processing, which is responsible for the chronic disease epidemic; and packaging with plastic and aluminium is responsible for nearly 50 per cent of all greenhouse gases emitted. Industrial agricultural production contributes 11-15 per cent of greenhouse gases; land use change and deforestation by agribusiness to grow GMO soya in the Amazon, and destroying the tropical rainforests to grow palm oil in Indonesia, Papua New Guinea, the Congo, and Paraguay is responsible for 15-20 per cent; and food waste contributes 2-4 per cent to climate-related problems and to hunger.

A linear extractive agriculture system is rupturing ecological processes and planetary boundaries, and violating human rights. The three boundaries at which we have already crossed safe limits are biodiversity integrity; genetic diversity; and the biochemical nitrogen and phosphorous cycles. All three are rooted in the chemical-intensive, fossil fuel-intensive model of agriculture. The erosion of genetic diversity and the transgression of the nitrogen boundary have crossed catastrophic levels.

Seventy-five per cent of the planetary destruction of soil, water, and biodiversity emanates from industrial agriculture which also contributes 75 per cent of food-related chronic diseases. The annual cost of chronic diseases is now in the trillions. Not only is the climate system being disrupted, biodiversity is disappearing. We could call the fossil fuel age an age of both climate disasters and species extinction: over 200 species are disappearing daily because of the assault by toxins on them.

Chemical agriculture does not return organic matter and fertility to the soil; it demands more water and destroys the soil's water-holding capacity. The first step to take in climate action is to change our way of thinking – from separation from the earth to non-separability, and from a dead earth paradigm to a living planet paradigm. In the process, we will activate the power to act, to co-create and co-produce with the earth, instead of being at war with her. We will move from a paradigm and economic system that creates scarcity to a paradigm and economic system that creates abundance and well-being for all humans and all species. The transition to regenerative organic agriculture is the single biggest climate action we can take in our communities and through our governments. It is capable of removing 100 per cent of greenhouse gases from the atmosphere, while also regenerating biodiversity, soil and water, and cultivating healthy and nutritious food.

Biodiversity-intensive, fossil fuel-free and poison-free agriculture produces more nutrition per acre while rejuvenating the planet. By intensifying biodiversity and following nature's law of return we can regenerate the soil-food web which is the source of recycling nutrients, including those nutrient cycles that connect soils and plants to the atmosphere. Since everything is interconnected and we are part of the earth, when we care for the earth, we also care for human well-being. It is the path to 'Zero Hunger' in times of climate change.

What governments need to do is to stop promoting the oil-based paradigm in schools and universities, and promote an earth-centred, soil-based way of thinking; they need to stop subsidising fossil fuels and industrial agriculture, and begin supporting communities that make a transition to local, ecological, biodiverse, poison-free and fossil fuel-free food and farming systems. *Earth care is climate action.*

Humus is the Latin word for soil; it is also the root of 'human'. Caring for the earth is an ethical and ecological duty. Through earthcare, we sow the seeds of our future, and of what I have called earth democracy. We also take action to avoid the certainty of climate catastrophe.

The year 2015 was the UN International Year of Soils. It was the year we witnessed multiple crises – desertification of the land, extreme drought due to climate change, and a massive displacement of people from their homes. We wanted to connect the degradation of the soil; the plight of refugees who were dying as their boats sank in the Mediterranean; and increasing conflicts like the one in Syria, which has become a war without end; as well as conflicts around the shrinking Lake Chad in Africa. I suggested to the ICFFA to organise a meeting that would connect the dots. I remember, while travelling in the train from Rome to Florence with my colleagues, Caroline Lockhart and Maria Grazia Mammuccini, I asked them for the Latin word for living earth and living soil. *Terra viva*, they said. The title of our Manifesto became *Terra Viva: Our Soil, Our Commons, Our Future*, and it was released at the Milan Food Expo in June 2015.

The same year we also undertook a soil pilgrimage in India that started at Gandhi's mud hut in Sevagram and ended at Albert Howard's laboratory in Indore. His book *An Agricultural Testament*, based on Indian farmers' indigenous knowledge, is the basis of the contemporary organic movement. Others who joined me in our soil pilgrimage were André Leu, president of IFOAM; Ronnie Cummins of the Organic Consumers Association; and 80-year-old Will Allen, one of the pioneers of the organic movement in the US.

In September 2015, I marched for the climate in New York with Bernie Sanders,[2] Bobby Kennedy, Jr., and Bill McKibben.[3] Those of us working on organic agriculture, biodiversity conservation and soil regeneration met in Costa Rica on Tom Newmark's[4] farm, Finca Luna Nueva, with others concerned about finding regenerative solutions to climate change. Regeneration International was launched with Hans Herren, chair of the UN International Assessment of Agricultural Knowledge, Science and Technology for Development. André Leu and Ronnie Cummins were on the founding steering committee, which soon expanded to include Precious Phiri[5] from the Africa Centre for Holistic Management, Ercilia Sahores[6] from Via Organica in Mexico, Renate Künast from the German Green Party, John Liu, the China-based filmmaker, and Tom Newmark and Larry Kopald from Carbon Underground. The same movements were present at the Paris Climate Change Conference to demonstrate that the culture of soil and earth care has the potential to find the answers to problems that the culture of oil has created. We joined hands to write a new pact for the earth.

Our survival demands that we make this new pact with the earth, and between diverse peoples, based on a new vision of planetary citizenship. A pact based on reciprocity, caring and respect, on taking and giving back, on sharing the resources of the world equitably among all living species. It begins by seeing and cherishing the soil as a living entity, a *terra viva*, whose survival is essential to our own.

The future will be cultivated from the soil and grow out of the land, not from the skewed global market of fictitious finance, corporate personhood, and consumerism. An earlier ecocentrism has given way to anthropocentrism which is now giving way to corporate-centrism. We need to move away from this corporocentric worldview to one centred on our earth family. Wherever we are on this planet, in all our diversity, the soil is our bedrock. We must, as earth citizens, reclaim it from corporate manipulation and greed, and care for it, together, in recognition of our common humanity and common responsibility.

On the eve of the 2015 United Nations Conference on Climate Change, the entire world looked towards Paris. This historic meeting was an invitation to all the peoples in the world to boldly make the shift from the paradigm of exploitation to one of gratitude and giving back; from privatisation and the enclosure of the commons to defending our commons of soil, seed, food, water and air. The climate crisis, food crisis, and water crisis are interconnected, and so are the solutions. They cannot be seen as separate.

In November 2015, before COP21, I was in Paris to launch the Organic Movement for the Greater Paris region, at the invitation of AMAP Network in Île-de-France, the organic movement linking farmers and consumers. On November 9, together with Clotilde Bato, who heads our partner in France, SOL (which means soil), I planted Gardens of Hope in the Jardin Marcotte and the Cultures en Herbes in Paris. During COP21, on the afternoon of World Soil Day – December 5 – leaders in defence of life and of our planet, representatives of Seed Savers (community supported agriculture networks), spiritual leaders, artists and concerned citizens gathered at Parc de la Villette in Paris and planted a Garden of Hope as a reminder that our seeds, our soils, and our biodiversity, kept in the hands of local farmers and caring citizens, are solutions to climate change. With the Garden of Hope and the new pact for the earth, we sowed the seeds of a new planetary citizenship. Our collective journey continues, in order to make peace with the earth.

Notes

[1] An organic dairy farmer in the UK, campaigner for sustainable food and farming, former head of the Soil Association, and founder of the Sustainable Food Trust.

[2] American politician and activist.

[3] American environmentalist, author, and journalist who has written extensively on the impact of global warming.

[4] Co-founder and chair of Carbon Underground; co-founder of Regeneration International; chairman of the Greenpeace Fund USA and the American Botanical Council; founder of Sacred Seeds, a plant conservation project; and co-owner of Finca Luna, a biodynamic and regenerative farming operation in Costa Rica.

[5] Training and development specialist on environmental issues, regenerative agricultural practices, and community organising.

[6] Political scientist specialising in international relations, adult education and community development, she is founding member, and Latin America director, of Regeneration International.

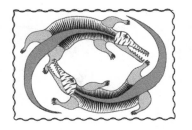

We are
the Biome,
We are
the Virome

WE ARE ONE EARTH FAMILY ON ONE PLANET, HEALTHY IN our diversity and interconnectedness. As Martin Luther King Jr. reminded us, 'We are caught in an inescapable network of mutuality, tied in a single garment of destiny. Whatever affects one directly, affects all indirectly.'

We can either be linked worldwide through the dispersal of a disease like COVID-19, when we invade the home environments of other species, manipulate plants and animals for commercial profit and greed, and spread monocultures; or we can be connected through health and well-being for all by protecting the diversity of ecosystems and the biodiversity, integrity, self-organisation (*autopoiesis*) of all living beings, including humans.

New diseases are coming into being because a globalised, industrialised, inefficient food and agriculture model is invading the ecological habitat of other species and manipulating animals and plants with no respect for their integrity. The health emergency that the corona virus is alerting us to is connected to another emergency, the extinction and disappearance of species, which, in turn, is connected to the climate emergency.

Over the past fifty years, 300 new pathogens have emerged, a result of the destruction of the habitat of species. According to the WHO, the Ebola virus moved from wild animals to humans, and *New Internationalist* reported that 'From 2014-16, an unprecedented Ebola epidemic killed more than 11,000 people across West Africa. Now scientists have linked the outbreak to rapid deforestation.'

Prof. John E. Fa of Manchester Metropolitan University, a senior research associate with the Center for International Forestry Research (CIFOR) says,

> Emerging diseases are linked to environmental alterations caused by humans. Humans are in much more contact with animals when you open up a forest.... You have a balance of animals, viruses and bacteria and you alter that when you open up a forest.

Take the Kyasanur Forest Disease (KFD). It is a highly pathogenic virus that spreads from monkeys to humans through virus-infected ticks, as deforestation shrinks the forest habitat of monkeys. The KFD virus is a pathogen that has long existed as part of an established ecosystem in South Kanara. Human modification of that ecosystem through deforestation caused an epidemic occurrence of the disease.

The corona virus, too, has come to us from bats. Author and journalist Sonia Shah observes, 'When we cut down the forests that the bats live in, they don't just go away, they come and live in the trees in our backyards and farms.' Prof. Dennis Carroll[1] of Cornell, too, acknowledges that as we penetrate deeper into ecozones that we did not occupy before, we create the potential for the spread of infection.

'Mad cow' disease, or bovine spongiform encephalopathy (BSE), is an infectious disease caused by deformed proteins called 'prions' that affect the brains of cattle. Cows were infected by mad cow disease when they were fed the rendered meat of dead, infected cows. When beef from the infected cows was consumed by humans, they were struck by

Creutzfeldt–Jakob disease (CJD). The prion is a self-infective agent, not a virus or bacteria.

Antibiotic resistance (ABR) is growing in humans because of the intensive use of chemicals in factory farms. Antibiotic resistance markers in GMOs could also be contributing to ABR. Horizontal gene transfer across species is a scientifically known phenomenon, which is why we have biosafety science and biosafety regulations like the Cartagena Biosafety Protocol, and national laws for biosafety. In the last few decades, non-communicable, chronic diseases have been spreading exponentially. A study in 2012 quantified the impact on health and costs related to damage resulting from exposure to 133 pesticides applied in 24 European countries in 2003, equal to almost 50 per cent of the total mass of pesticides applied in that year. The findings revealed that only 13 substances applied to three classes of crops (grapevines, fruit trees and vegetables) contributed up to 90 per cent of the overall health impacts due to a loss of about 2,000 years of life (corrected for disability) in Europe every year, corresponding to an annual economic cost of €78 million. Another 2012 survey that assessed the cost of acute pesticide poisoning in the state of Paraná in Brazil concluded that the total amounted to US $149 million each year. That is to say, for every dollar spent on the purchase of pesticides in this state, around US $1.28 was spent due to costs externalised by poisoning.

It has been estimated that in the 1990s, in the United States, the environmental and public health costs resulting from the use of pesticides amounted to US $8.1 billion every year. Therefore, US $4 billion were being spent every year for pesticide consumption in the country; for one dollar spent on the purchase of these substances, two dollars were for outsourced costs. Another study published in 2005 estimated that in the US, the costs of chronic diseases through pesticide poisoning amounted to US $1.1 billion, of which about 80 per cent was accounted for by cancer. It has also been calculated that in the Philippines, the transition from one to two pesticide treatments for rice cultivation resulted in an increased profit of 492 pesos, but created additional health costs of 765 pesos, with a net loss of 273 pesos. In Thailand, it has been estimated that the externalised cost of pesticides can vary annually from US $18 to 241 million. In Brazil, the costs for damage to the health of workers employed in bean and maize crops amounts to 25 per cent of the profits.

With regard to data which is closer to the European reality, a study by a panel of experts to assess the burden of diseases and costs related to exposure to endocrine disruptors evaluated, with 'strong probability', that every year in Europe, about 13 million lost IQ points and 59,300 additional cases of intellectual disability could be attributed to organophosphate exposure. Since it has been estimated that each lost IQ point for prenatal exposure to mercury is worth about €17,000, similar accounts can be rendered for exposure to organophosphorus.

The health consequences of maladapted modernity are being experienced in epidemic proportions across the world. Apart from premature death and prolonged disability, diseases resulting from nutritionally deficient diets are forcing people to seek expensive, often unaffordable, healthcare. Commercial healthcare systems are the beneficiaries of these modern epidemics, offering technology-intensive and high-cost tests and treatments for health disorders that could, and should, have been easily prevented through good nutrition and a healthy environment. The merger of Bayer and Monsanto implies that the corporations that sell the chemicals that cause diseases also sell the pharmaceutical cures for the diseases they have caused.

The global costs of healthcare due to food-system related illness are

- Obesity: US $1.2 trillion by 2025.
- The global cost of just diabetes in 2015 was estimated at US $1.31 trillion. In Italy, every patient suffering from diabetes today costs €2,589 annually to the National Health System, and diabetes-related therapies comprise roughly 9 per cent of its budget, or about €8.26 billion. In Africa, 35 million people – twice the number at present – will be affected by diabetes, and by 2030 this is estimated to cost US $1.5 trillion.
- Antimicrobial Resistance (AMR) infections: US $1 trillion by 2050.
- Cancer: US $2.5 trillion.
- Cost of exposure to endocrine disruptors in Europe: US $209 billion annually; the cost of exposure to endocrine disruptors in the US: $340 billion.
- New research indicates that in the US, the annual cost of autism has more than tripled to US $126 billion; autism costs touched £34 billion in the UK, and is its most expensive health problem.
- Rising infertility has led to a new fertility industry which cost the US $21 billion in 2020.

Governments need to ensure that biosafety and food safety assessments are not influenced by the industry that benefits from manipulating living organisms and suppressing scientific evidence of harm. The harm caused to people's health by the corporate manipulation of research has now been proven. The global attempt at deregulating food safety and biosafety requirements must be stopped.

With the corona virus, governments have shown that they can take action to protect the health of people if they have the will to do so. The COVID crisis and the response to it need to become the basis for halting those processes that degenerate our health and the planet's health, and renewing those processes that regenerate both. The health emergency has shown that the right to health is a fundamental right, that health is a commons and a public good, and that governments have a duty to protect public health. Health is a continuum, from the soil to plants, to our gut microbiome. The future depends on our oneness as humanity on one planet, connected through biodiversity and health. Let us not allow the cautions of today to be cemented into a permanent climate of fear and isolation. We need each other and the earth, in our rich diversity and self-organisation, to create resilience in times of emergency, and to regenerate health and well-being in the post-corona world. This crisis has created a new opportunity to effect a paradigm shift from the mechanistic, industrial age of separation, domination, greed and disease, to the age of Gaia, of a planetary civilisation based on the consciousness that we are one earth family, that our health is one health rooted in ecological interconnectedness, diversity, regeneration, and harmony.

The human microbiome consists of all the microbes — bacteria, fungi and viruses — that live within us or on us, including the skin, mammary glands, seminal fluid, uterus, ovarian follicles, lungs, saliva, oral mucosa, conjunctiva, biliary tract, and gastrointestinal tract. It has been estimated that there are over 380 trillion viruses inhabiting us, a community collectively known as the human virome. More than 38 trillion bacteria are a part of us, the human biome. Our gut microbiome has 100 trillion micro-organisms and 100 species. There are 100,000 times more microbes in our gut than people on the planet. There is an intimate connection between the biodiversity and health of soils and plants, and our gut and our brain. Our gut is a microbiome that contains trillions of bacteria. To function in a healthy way, the gut microbiome needs a diverse diet, and a diverse diet needs diversity in our fields and gardens. The loss of diversity in our diet creates ill health. Over the last decade, western science has begun to acknowledge the centrality of the gut to health, a principle that ayurveda followed over ten centuries ago. As Emeran Mayer's *The Mind-Gut Connection* acknowledges,

> For decades, the mechanistic, militaristic disease model set the agenda for medical research: As long as you could fix the affected mechanical part, we thought, the problem would be solved; there was no need to understand its ultimate cause…. We are just beginning to realise that the gut, the microbes living in it — the gut microbiota — and the signalling molecules that they produce from their vast number of genes — the microbiome — constitute one of the major components of these regulatory systems.

The gut is increasingly being referred to as the second brain. It has its own nervous system, the enteric nervous system, ENS, with 50-100 million nerve cells. Our bodies are intelligent organisms, and intelligence is not localised in the brain. A healthy gut has diversity as well as an effective barrier between the inside of the gut, where microbes break down the food, and the gut-associated immune system which allows healthy interaction and communication between gut microbes and immune cells. The greater the biodiversity in any ecosystem, the greater its resilience and resistance to disease. This also applies to our gut ecosystem. Biodiversity destruction in the gut microbiome is responsible for inflammation and metabolic dysregulation leading to many chronic diseases, including Type 2 diabetes, obesity, cognitive decline, depression and degenerative brain disorders. When our gut biodiversity plummets because of the toxicity or deficiencies in the food we eat, health pandemics can emerge – gastrointestinal infections; autoimmune diseases like asthma; rheumatoid arthritis; inflammatory bowel disease; autism spectrum disorders; obesity; and metabolic diseases.

'One health' implies that we recognise that microbial biodiversity is essential for building our immune system. Just as soil microbes help plants to grow and stay healthy, the microbes in our body provide us with nutrients and maintain our physical and mental health. They strengthen our resilience when facing disease. Since we are more bacteria than human, when the poisons we use in agriculture reach our gut through food, they can kill beneficial bacteria. Our gut microbes process the food we eat and transform it into nourishment for our body and brain. Our gut microbiome biodiversity performs the vital ecological functions of providing and absorbing nutrients, protecting against pathogens, maintaining the barriers that filter what is beneficial and what is harmful to health, and transforming our food into the diverse chemicals and enzymes that maintain it. The gut microbiome participates in vital processes, including digestion, energy homeostasis and metabolism, the synthesis of vitamins and other nutrients, and the development and regulation of immune function. It also contributes to the production of numerous compounds that enter the bloodstream and affect various tissues and organs of the body.

The bacteria in our gut produce three aromatic amino acids, tryptophan, tyrosine, and phenylalanine, through the shikimate pathway. Since our cells don't have this pathway, they themselves are unable to make these nutrients; we depend on our gut bacteria to do so. These essential amino acids are precursors to the neurotransmitters dopamine, serotonin, melatonin and adrenaline, as well as the thyroid hormone, folate and vitamin E. The killing of gut bacteria leads to deficiencies in these important biological molecules, and impairs our neurological functions. The Centers for Disease Control (CDC) data show that, based on current trends, one-in-two children in the US will show signs of autism in the next few decades.

Ayurveda recommends six tastes for a balanced diet – sweet, sour, salty, pungent, hot, bitter, and astringent. Each taste carries the potential for processes that create and sustain the self-regulating systems of our body. Taste receptors do not just lie in the tongue, they are distributed throughout the gastrointestinal tract and are located on sensory nerve endings and on the hormones containing transducer cells in the gut wall. New biological science is now finding that the gut has sensors for different tastes, and different metabolic processes are governed by the diversity of tastes; for example, there are 25 different bitter taste receptors.

Specific molecules and phytochemicals found in herbs and spices activate specific taste receptors and trigger particular metabolic processes. Sweet receptors stimulate the absorption of glucose into the bloodstream and the release of insulin from the pancreas. Mayer states in *The Mind-Gut Connection*,

> The multitude of phytochemicals derived from a diet rich in diverse plants, combined with the array of perfectly matching sensory mechanisms in our gut, synchronises our internal ecosystem, our gut microbiome with the world around us.... The gut's sensory systems are the National Security Agency of the human body, gathering information from all areas of the digestive system, including the esophagus, stomach and intestine, ignoring the great majority of signals but triggering alarm when something looks suspicious or goes wrong. As it turns out, it's one of the most complex sensory organs of the body.

Eating is a conversation between the soil, the plants, the cells in our gut, and the cells in our food, and between our gut and brain. Eating is an intelligent act at the deepest cellular and microbial level, and cellular communication is the basis of health and well-being. It is also the root of disease. Poisoned food creates disease. We might be ignorant about the links between food and health, but our cells know them. The US, the most 'advanced' country for the industrialisation of food and agriculture, has 40 per cent less gut microbiomes in its population than indigenous communities in the Amazon and Africa. The gut biodiversity of those living pastoral and agrarian lives is much richer than that of those in industrial society. When it comes to health, which depends on biodiversity in the gut, communities considered 'backward' are much more advanced than those considered 'advanced'. Grains like millets, called 'backward' and 'primitive', are the better food for the health of people and the planet.

Scientists have only begun studying the human virome and human biome over the last decade or so, yet they are ready to 'conquer' the 'bad viruses' and create humans as 'super organisms', a continuation of anthropocentrism and a militaristic epistemology. On December 1, 2020, David Pride, an infectious-disease specialist at the University of California, wrote in *Scientific American*, 'If we humans can figure out how to manage the bad viruses and exploit the good ones, we could help ourselves become stronger super organisms.' However, it is precisely such an attitude of control and conquest that is at the root of disease pandemics. There are no good or bad viruses in an absolutist, essentialised, atomised, deterministic sense. A virus that is safe in one ecosystem, and as part of one organism, can become unsafe for other organisms when it is displaced. The corona virus in bats is safe for the bats; the monkey virus was safe for the monkeys in the forest. When humans invaded their homes and habitats and displaced animals, the viruses jumped from animals to humans and became unsafe, spreading disease epidemics. The COVID-19 experience should be a lesson in humility, a call to recognise our interconnectedness with the rest of the life forms, to not cause harm to other beings and species which translates into harm to ourselves.

It is the science of biodiversity and interconnectedness between living soil, living seed, living food, living economies of well-being that Navdanya has been practising and promoting for the last three decades. For us, earth care is healthcare, of people and the planet.

Notes

[1] A senior public health expert, he serves as chair of the Global Virome Project Leadership Board, and was previously director of USAID's Pandemic Influenza and Other Emerging Threats Unit.